DECOLONISATION AND CRITICISM

CONTEMPORARY IRISH STUDIES

Series Editor Peter Shirlow (School of Environmental Studies, University of Ulster, Coleraine)

Also available

James Anderson and James Goodman (eds)
Dis/Agreeing Ireland
Contexts, Obstacles, Hopes

Colin Coulter
Contemporary Northern Irish Society
An Introduction

Graham Ellison and Jim Smyth
The Crowned Harp
Policing Northern Ireland

Marie-Therese Fay, Mike Morrissey and Marie Smyth
Northern Ireland's Troubles
The Human Costs

Paul Hainsworth (ed.)
Divided Society
Ethnic Minorities and Racism in Northern Ireland

Denis O'Hearn
Inside the Celtic Tiger
The Irish Economy and the Asian Model

Peter Shirlow and Mark McGovern (eds)
Who are 'the People'?
Unionism, Protestantism and Loyalism in Northern Ireland

Gerry Smyth
Decolonisation and Criticism
The Construction of Irish Literature

Gerry Smyth
The Novel and the Nation
Studies in the New Irish Fiction

DECOLONISATION AND CRITICISM

The Construction of Irish Literature

Gerry Smyth

LONDON • ANN ARBOR, MI

First published 1998 by Pluto Press
345 Archway Road, London N6 5AA

British Library Cataloguing in Publication Data
A catalogue record for this book is available from the British Library

ISBN 9780745312323 hbk
ISBN 0745312322 hbk
ISBN 9780745312279 pbk
ISBN 0745312276 pbk

Library of Congress Cataloging in Publication Data
Smyth, Gerry, 1961–
Decolonisation and criticism: the construction of Irish literature/Gerry Smyth.
p. cm. — (Contemporary Irish studies)
Includes bibliographical references and index.
ISBN 0-7453-1232-2 (hbk)
1. English literature—Irish authors—History and criticism—Theory, etc. 2. Literature and history—Ireland—History—20th century. 3. National characteristics, Irish, in literature.
4. Ireland—Intellectual life—20th century. 5. Criticism—Ireland—History—20th century. 6. Decolonization in literature. 7. Group identity in literature. 8. Colonies in literature. I. Title.
II. Series.
PR8718.S69 1998
820.9'9417—dc21 98-3719
CIP

Designed and produced for Pluto Press by
Chase Production Services, Chadlington, OX7 3LN
Typeset from disk by Stanford DTP Services, Northampton
Printed & bound by Antony Rowe Ltd, Eastbourne

For my mother

Contents

Preface

Like every other field of modern intellectual endeavour, Irish Studies reveals a stand-off between traditionalists and those – many? most? – who are interested in extending the field or challenging some of its base assumptions. Even a brief glance reveals that most of the institutional power lies in the hands of the former and that Irish Studies remains for the most part locked within conservative modes of scholarly analysis. Traditionalists accuse radicals of being mesmerised by fashionable antihumanist philosophies which have made their way into reputable disciplines such as historical and literary studies. The infiltration of Irish Studies by poststructuralist and postcolonialist modes of thought threatens (so it goes) to damage the credibility of the entire field, with jargon-laden monographs, thin research and sloppy scholarship the inevitable results. As a consequence, many of those with their hands on institutional power consider it a duty to exclude or ignore any initiative which challenges the *status quo* whenever and wherever it emerges.

It is also clear that most British and American (and many Irish) publishing houses contribute significantly to the conservatism of Irish Studies. The major players appear happy to remain within the received limits of Irish literary and historical debate. Houses enjoying a 'radical' reputation in fields such as Gender, Cultural Studies, Ecocriticism, etc., have shown a remarkable reluctance to entertain any conceptual and/or methodological departure from established practices when the subject concerns Ireland. It may be that British and American editors feel that the radical gesture lies in the fact of *having* an Irish Studies portfolio at all, irrespective of its constituency or its constitution. In this way the 'debates' are eked out: another combination of the usual suspects confronted with another combination of the usual questions.

While obviously caught up in such questions, this book is also about them. It is thus from a quaking sod – simultaneously *a part of* and *apart from* its subject matter – that *Decolonisation and Criticism* sets off into the world. The 1950s are unfashionable enough in Irish Studies, a sort of Dark Ages between the Classicism of cultural nationalism and the Renaissance of post-nationalism. But this would not be enough by itself to constitute a challenge to established practices, as we know that researchers are routinely encouraged to colonise new periods and marginal subjects in the drive to extend the discursive remit (and the market) of this or that field. My entreaty to the contemporary Irish Studies community is more fundamental, namely, that it should attend to its own role in the organisation and dissemination of narratives of power. Accordingly, in this book all the assumptions rendered invisible in traditional practices – including the key areas of publishing, criticism and consumption – are interrogated for the ideological work they have performed and continue to perform in a range of sociopolitical contexts. Thus, although the analysis of critical material in Part 2 stops around 1958, every time an interpretation is offered of a critical text from the 1950s it is with self-conscious reference to subsequent and current debates. The text is not saying: 'Here we are, critic, publisher and reader in 1998, above and beyond all *that*, contemplating how it *really* was from our privileged position forty years hence.' It is saying: 'Every claim this text makes about the critical discourses of the past is informed with the assumptions of the present; at the same time, those assumptions are being interrogated by the critical discourses and practices of the past.' It is not the past which is on trial here; it is the present and what we as critics, publishers and readers do in the present in the name of the past.

Decolonisation and Criticism may well be attacked by those who have an interest in maintaining the status quo, be they publisher, academic-lecturer or intellectual-ideologue. I hope it will be welcomed by those who are frustrated with the limitations of the current dispensation and are searching for new directions in which to pursue Irish-related research.

Earlier versions of Chapters 1 and 2 have appeared as: 'Decolonisation and criticism: towards a theory of Irish critical discourse' in *The Mechanics of Authenticity: Contemporary Irish Cultural Criticism*, edited by Colin Graham and Richard Kirkland (Basingstoke: Macmillan, 1998); and 'Writing about writing and national identity', in *Contemporary Writing and National Identity*, edited by Tracey Hill and William Hughes (Bath: Sulis Press, 1995) 8–17.

I would like to thank the following people for their advice, support and inspiration: Timothy Ashplant, David Cairns, Peter Childs, Colin Graham, Matt Jordan, Declan Kiberd, Kim Perkins, Mike Pudlo, Neil Sammels, Joan Smyth, Sheila Smyth, Kevin Smyth, Roger van Zwanenberg, the South Roadies. Anyone who knows Shaun Richards will also know how fortunate I was to be able to work with him during the early stages of this study. He is a scholar and a gentleman. Love and thanks as always to my family: Christine, Lizzie and Esther.

Gerry Smyth
West Kirby
February 1998

Introductions

This book investigates the role played by the discourse of literary criticism in the process of Irish decolonisation since the late eighteenth century, with special emphasis on the 1950s. It is not *about* Irish literature as such – no new readings of the usual suspects – but about the ways in which Irish literature has been constructed in the modern era as a specific realm of cultural activity possessing particular effects and properties. *Decolonisation and Criticism* looks to reconfigure the established relations between (primary) literature and (secondary) criticism, and then to set this analysis against a modular theory of decolonisation based on a reading of Irish history from the perspectives of contemporary postcolonial and post-structuralist theory. The text is structured into two main parts. In Part 1 I shall argue that dominant anti-colonial practices are always ultimately undone by operating uncritically within the discursive and psychological limits set by colonial discourse. This argument will proceed by way of a discussion of theories of decolonisation, especially as they pertain to Ireland (Chapter 1); an analysis of the function of literary criticism as it relates to decolonisation (Chapter 2); and a brief history of the emergence of an Irish critical tradition which posited a specific and crucial relationship between literature and nation (Chapter 3).

In Part 2 attention will be focused on the possibilities for a 'post-colonial' Irish identity as mediated by the discourse of literary criticism. Concentrating on a short period, and by way of close analyses of a number of critical practices and interventions (Chapters 5 to 7), it will be argued that the appellation 'post' is premature in this context, and that Irish critical discourse in the period after 1922 remained trapped with the discursive practices and psychological structures of the former (anti-) colonial period. This was also the period, however, in which many Irish people began to reflect seriously and at length upon the historical narratives in which they were caught up. I shall be arguing that 'postcolonial' Ireland's most radical and self-conscious thought emerged initially in terms of literary critical discourses, and that the most challenging critical work of the period between 1948 and 1958 is characterised by the struggle to imagine ways of escaping the discursive economy inherited from the eighteenth and nineteenth centuries.

This is as much as I wish to say about the structure and thesis of *Decolonisation and Criticism* at this stage. As I understand it, the usual thing in an 'Introduction' is for the author to position himself in relation to current debates, to describe the methodology underpinning the analyses, to define/defend the project's structure and limits, etc. The preceding two paragraphs have made a gesture towards that established practice. However, books about criticism (such as this one) are unusual in that a normally implicit relationship between subject matter (what the book is about) and methodology (how the material is organised and addressed) is laid bare. There have been many moments in this book when I have found myself addressing some example of critical discourse from Irish history only to discover that it is I who am in fact being positioned and addressed by that voice from the past. Put another way: there is always a danger in books of this kind that *what* the critic says about literary criticism (content) will rebound on his own critical practice, that is, on *how* he says it (form).

This is not just a modish poststructuralist paradox. It *is* a recurring paradox which attends all critical discourse (but especially *meta-criticism* such as this), one with specific intellectual, personal and political ramifications which the more *engagé* forms of poststructuralism are endowed to reveal. Consider first of all the theoretical and methodological bases for this study. They are for the most part responses to developments that have occurred within Irish Studies and in wider fields of cultural and critical theory since, roughly speaking, the 1960s. These concerns have determined my archive of material and the means by which I address it. On the one hand, this includes institutional and wider critical developments such as the growth and concomitant crisis of Irish Studies in Great Britain, the various initiatives grouped together under Field Day, the revisionism debate; this book also bears the marks – explicit and implicit – of specific textual interventions such as *Writing Ireland* by Cairns and Richards, Lloyd's *Anomalous States* and Kiberd's *Inventing Ireland.* In terms of modern cultural/critical theory, on the other hand, it involves the impact of francophone research on Anglo-American intellectual discourse, the emergence of 'cultural studies' and 'postcolonialism' as fashionable intellectual paradigms, and the constitution of 'theory' as an area of debate in and of itself. Here, texts such as Said's *Orientalism*, Derrida's *Of Grammatology* and Bhabha's *The Location of Culture* constitute significant moments.

Such methodological and theoretical eclecticism is a contentious issue within contemporary Irish critical debates. An established manoeuvre is for the leading critics to reflect (frequently in Introductions to essay collections and monographs of the sort you are now reading) upon the possibility and the constitution of an Irish literary criticism. For example, a variety of positions (albeit unstable and shifting) regarding the mixture of Irish literary history and modern Anglo-American theory are available depending on whether you read W.J. McCormack or Seamus Deane, Edna Longley or Luke Gibbons, David Lloyd or Roy Foster.[1] Introductions clear the theoretical and methodological ground, laying the basis for what follows. I could likewise rehearse these positions here and locate my work in relation to them, expanding upon the schema of the previous paragraph before going on to practice (in the 'real' book, Chapters 1 to 7) what I have preached in this Introduction. In some ways that is what I am in fact doing right now. The point is this, however: this debate ('the possibility and constitution of an Irish literary criticism') bears not only upon the *form* of *Decolonisation and Criticism*: *it is the text's specific subject.* Therefore, I have frequently found that when analysing a critical text from the past (as part of the content), such analyses took on a self-conscious or reflexive dimension, such that it was my own critical practice in the present (as part of the form) which became the subject. *Decolonisation and Criticism* thus is a mirror or an echo of itself, both form and content, both methodology and archive.

For example, in 1956 Vivian Mercier published a short article entitled 'An Irish School for Criticism?' in the prestigious journal *Studies*. This was a reply to two previous essays (by Denis Donoghue and Donald Davie) regarding the possibility and the constitution of an Irish literary criticism. In a brief argument, Mercier demonstrates his conversance with developments in America and Great Britain, deplores the absence of a native critical tradition and finally calls for a full scholarly programme, irrespective of methodology, to address Irish literary history. It is interesting to me that Irish (and English) intellectuals of the 1950s were self-consciously deliberating upon the question of Irish criticism; that they were doing so in the pages of a prestigious though conservative journal; that they showed familiarity with international developments (such as American New Criticism) and were engaged with considering the relations between those developments and indigenous initiatives; and, most crucially, that they linked the fate of critical discourse in Ireland with wider historical and political discourses bearing upon the state of the nation. At the same time,

Mercier engages with many of the issues (concerning methodology and archive) which have influenced the organisation of this book. It is thus also interesting to me, for example, that subsequent criticism has served to marginalise the Irish writers and critics of that interim period between the heady pre-revolutionary days and what is to all intents and purposes another, third, revival of Irish cultural activity. The issues with which Mercier grappled back in 1956 speak forward to many of the same issues with which I and other Irish critics grapple as the millennium approaches. When addressing 'An Irish School for Criticism?', therefore, I find myself constantly slipping between interpretative strategies, reading now for what it reveals about the function of Irish literary criticism during the 1950s, and now for how it informs and engages with current debates, especially my own fix on those debates. In the light of what I have to say in Chapter 2 regarding the relations between criticism and decolonisation, this was an interesting revelation.

Recourse to personal history is something else that is at odds with standard critical discourse. Again, however, metacriticism puts pressure on all the 'standard' effects of that form of institutional literary criticism which is still the dominant paradigm in late twentieth-century academic practice. The dual intellectual provenance described above stems initially from personal circumstances: I am an Irishman interested in Irish history and literature but trained in certain methodological techniques and cultural assumptions at a number of English universities. Historicising critical discourse from the past compels one to historicise one's own asumptions. Detailing the particular crises and debates encountered by dead critics leads the contemporary critic to an understanding of how current practices are in dialogue with those same crises and debates, whether they know it or not. There are practical as well as philosophical implications to this. Two of the more hard-line Irish literary critics mentioned above have in recent years called for (and practiced) a more self-conscious approach to criticism. Referring to Irish higher educational practices, Longley writes: 'Research that historicises our own activities can help us to walk the wobbly tightrope between attachment and detachment' (1996/97: 5); while in the course of some serious Theory-bashing, McCormack has insisted (in terms reminiscent of both Foucault and Derrida, as he must know) that 'One must work from one's actual locus and not some universalised ideal professional vantage point' (1993: x). If nothing else, then, *Decolonisation and Criticism* might provide contemporary Irish literary critics (especially those dazzled by their own originality) with a genealogical sense, reminding them that discourse, especially critical discourse, is intertextual and interested rather than detached and disinterested.

All told, a series of relationships might be said to inform *Decolonisation and Criticism*, emerging eventually as a recognisable if implicit relationship between *content* and *form*: Irish background *and* English education; Irish Studies *and* postcolonial theory; 'criticism' (*what* the book is about) *and* 'criticism' (*how* it is written), past (pre-1960) *and* present (post-1960). But consider the structure of *Decolonisation and Criticism* as described above: the material (that is to say, the 'content') surveyed in Part 2 ends around 1960. In historical terms, therefore, the 'content' (pre-1960) comes *before* the 'form' (post-1960). Yet at the same time, it is the 'form' (my engagement with modern critical theory and Irish Studies since the 1960s) which precedes (Part 1) and makes possible the 'content' (Part 2, the analysis of pre-1960 Irish literary criticism). This means that Part 2 is a pre-history of Part 1; the subject matter feeds *into* rather that *out of* the form. Much of the time, the arguments with which I engage in Part 1 are themselves tacit engagements with (when they are not in fact deliberate rejoinders to) arguments from before 1960 (Part 2). This also has the effect that any standard 'Introduction' (describing the text's theoretical and methodological premises in relation to current debates) would also be a 'Conclusion' in as much as it took up (Irish criticism since 1960) where the subject matter (Irish criticism up to 1960) leaves off.[2] (At this stage readers are invited to peruse the Table of Contents or the 'Notes' towards the end of the book to assure themselves that there is, so to speak, a text in this book, and that they do not have to plough through 200 pages of such tortuous argumentation.)

For many, this is the stuff of Borgian (or Derridean) nightmare, abandoning us to a world of diminishing reality in which books are written about books, Theory triumphs over history, and primary causes such as meaning and intention are relegated to the status of secondary effects. The problem may not be so much with criticism, however, as with the sorts of things that we have been led to expect from criticism and, more generally, from research. The dream of a stable self-conscious knowledge community conversing within a stable accepted language has long since dissipated; it was never much of a dream anyway, and history (yes, history) shows that the most exciting scholarly work has always emerged from challenges to methodological and archival orthodoxies. But given the exponential rate at which information is amassing as I write and as you read, even the very possibilities of 'debate' and 'intervention' may be receding. The fact is that irrespective of (rather than because of) my intentions or your desires, reader, *Decolonisation and Criticism* seems bound by dint of its very existence to contribute to 'the expansion of uncertainty' (Barnett, 1997: 18) in our field and in the arena of knowledge generally. It is my hope, however, that it may also contribute to the process whereby we come to cope with uncertainty rather than fearing or attempting to out-manoeuvre it.

These issues are dealt with at some length in Chapter 2 and are implicit throughout the text. In the meantime, an analogy offers itself from the uncertain relations between James Joyce (in many ways the silent hero of this book) and the character of Stephen Dedalus in *A Portrait of the Artist as a Young Man*. In that novel Joyce wrote a story, as Umberto Eco puts it, 'of a young man who wants to write *A Portrait*' (1982: 10). Just as Stephen/Joyce lives through what he will one day formalise into art (the end of the story, the beginning of writing), so this book about criticism is a description of its own possibility, a description of the critical past that has fed into and engaged with the critical present (and the act of description). The point? Simply this: *Decolonisation and Criticism* is quite obviously an articulation of *my* training, beliefs and prejudices; but it also relates part of *your* history, in as much as we shall be tracking in the next seven chapters a tradition which has enabled the emergence of a range of critical effects: this text, the subject who wrote it and the subjects who might read it.

PART 1

DECOLONISATION AND CRITICISM

1 The Modes of Decolonisation

The most contentious debate in contemporary Irish Studies concerns the establishment of the proper basis upon which to address the political and cultural activity of the modern period.[1] We may assume that most of the people who engaged in political and/or cultural activity over, say, the past 250 years had their own motives and desires regarding the kind of Ireland in whose name they acted and which they worked to realise. However, at the same time as contemporary critics describe the manifestation of such motives and desires in political and/or cultural phenomena, they also engage (whether acknowledging it or not) with the validity of those same motives and desires. When Yeats wrote that he desired to be accounted one who sang 'to sweeten Ireland's wrong', any critical response to the writing also functions as a response – however tacit, however remote – to Yeats's very particular understanding of Irish history. That is to say: critical analyses of texts always include an implicit critique of the (political and cultural) motives and desires underpinning texts. This is why so much attention continues to be devoted in Irish Studies to beginnings, to establishing the first principles of Irish history in terms of which the motives and desires of political/cultural actors may be addressed.

The assumption upon which this study is based is that a colonial relationship obtained between Ireland and England since the twelfth century, that such a relationship underwent simultaneous consolidation and crisis during the eighteenth century, and that modern Irish political and cultural activity may as a consequence most usefully be understood in terms of a model of *decolonisation.* While it seems clear that the Irish experience of colonialism was unique, this is not to subscribe to the myth of an ideal or stable or 'real' colonialism against which an aberrant Irish history must be measured. I am in fact concerned with colonialism in this book only in so far as it was perceived and resisted by certain kinds of Irish subjects. But as a *first* principle I shall assert that 'colonialism' continues to be a useful *general* term (that is, in Ireland and elsewhere) in that it encompasses a cluster of political and cultural discourses characterised by a precarious discursive economy of identity and difference (of which more below). As a *second* principle

I shall assert that the history of English colonialism in Ireland is *exemplary* in the degree to which it was unstable, inconsistent and partial. A corollary is that Irish history reveals a narrative of decolonisation which has also been unstable, inconsistent and partial. This in turn has led to the multitude of historical responses to the fact of colonialism as well as to the constant (and ongoing) questioning of the propriety of the colonial model itself. Nevertheless, it is from this complex matrix of the *general*, the *unique* and the *exemplary* that the assumptions informing *Decolonisation and Criticism* spring.

In this first chapter I wish to suggest that a flexible, non-deterministic model of decolonisation, encompassing a number of specific 'modes', enables us to account for the wide range of responses since the eighteenth century to the question of Ireland's identity. More subtly, such a model also enables us to engage with the salience of the category of 'identity' itself as the terrain upon which the vast majority of political, cultural and critical encounters in Irish history have taken place. In deploying such a model we shall see how nationalism, for example, came to constitute a significant strategy at a particular moment in the history of Irish, Anglo-Irish and British relations. We would also be obliged to acknowledge, however, that nationalism emerged and moved to a position of dominance in Irish political and cultural history only in competition with other narratives and other possibilities. I now wish to describe at a more or less general level what I understand to be the political and philosophical dynamics of this model, before going on in the final section to look at how versions of it have been deployed in Irish Studies in recent years.

Nationalism and Decolonisation

How does a colonial community set about imagining and organising a discourse of resistance? Theories of colonial resistance deriving from a wide range of methodological, disciplinary and political sources have tended to focus upon two main and related issues: a fundamental doubleness at the root of Western nationalism, and an unavoidable derivativeness underpinning Second- or Third-World anti-colonialism. Brought together in the analysis of decolonising formations, such theories represent a dualistic model, both elements of which – nationalism *and* anti-colonialism – are characterised by crisis. Decolonisation, that is, is subject on the one hand to nationalism's ambiguous (dualistic) nature while at the same time revealing the extent to which colonial resistance is fatally bound to the Western imperialist systems against which it is obliged to cast itself. Let us track the emergence of this model and its twin dilemmas in more detail.

In the first place, many commentators have claimed to discover a fatal division at the heart of modern European nationalism. This idea characterises many of the traditional studies of nationalism, and informs what has come to be seen as its 'invented' status.[2] One variation sees nationalism as 'the modern Janus' (Nairn, 1981: 329–63), a concept split between what Anthony Giddens describes as 'virulent forms of national aggressiveness, on the one hand, and democratic ideals of enlightenment on the other' (1987: 218). Born of the same changes in the structure of European thought as those which ushered in the Enlightenment, nationalism is believed to reveal those roots precisely in the readiness with which it has been turned to both emancipatory and despotic ends.[3]

Tzvetan Todorov has refined the division between 'good' and 'bad' nationalisms in two significant ways; by identifying, first of all, a split between *ethnic* and *political* entities, and secondly, a split between *internal* and *external* orientations (1993: 171, 175). In Todorov's discourse, moreover, both these divisions are overwritten by an ethical imperative in which a benign cultural universalism is cast against a malign political chauvinism. He understands the modern nation as an ongoing amalgam of the cultural and civic discourses that are basic to human existence, discourses which are historically and theoretically distinguishable but which took an inexorable turn towards identification during the eighteenth century. Cultural nationalism (based on custom, language and communal memory) is opposed to civic nationalism (based on citizenship, social rights and obligations). During the eighteenth century, this division evolved into a split between an *internal* nationalism (the *equality* of all national subjects cast in the language of popular sovereignty) and *external* nationalism (the patent *inequality* of different nations). Categorically divisible, these two tendencies were fatally fused under the influence of American and French Revolutions, the first (culturalist/universalist) gradually being subsumed by the second (political/anti-universalist) so that since the nineteenth century the modern world has been organised around the accepted coincidence of cultural and political entities. Todorov is in no doubt, moreover, about the malign influence of this confluence between cultural and civic nationalism in world history:

> At the end of this process, nationalism indeed appears to bear the principal ideological responsibility both for the European wars, from the Revolution up to and including the First World War, and for the colonial wars of the same period and beyond. Even if a war has other than ideological causes, we may attribute to these doctrines, without fear of contradiction, the responsibility for the deaths of millions of human beings and for political situations whose resolution, in many instances, is not yet in sight. (263)

We find many similar resonances in the work of another major theorist of nationalism, Anthony D. Smith. Like Todorov, Smith distinguishes between 'cultural' and 'civic' nationalisms, discovering the roots of these variants in the eighteenth century.[4] Unlike Todorov, however, he is at pains to stress the ethically uncertain orientation of modern nationalism, its ability to be wielded for and from a wide range of ideological positions. Smith's point of departure is a perceived crisis amongst European intellectuals of the eighteenth century, forced to address a major gap that appeared between traditional forms of legitimisation based on received religion and tradition on the one hand, and a rationalist project organised around the principles of reason, autonomy and observation, on the other. Both 'ethno-nationalism' and 'civic nationalism' are attempts to address this crisis, the former by providing a form of legitimisation based on the authenticity and longevity (and ultimately the mystique) of the *ethnie*, the latter by appeals to the autonomous, rational state as the basis of civilisation.

Emerging from existing concepts of 'national character' and 'national genius', ethno-nationalism is organised around culturalist tendencies (primarily medieval and ruralist in orientation) towards history, folk customs and rituals, and landscape analysis. It is concerned to 'vernacularize' (Smith, 1991: 140) the masses, to render the *ethnie* a 'subject of history' (126) as a prelude to formalising that subject into a state. For Smith, this form of nationalism is by and large an invention of intellectuals and artists concerned to provide the *ethnie* with a historical rationale and to create a language capable of disclosing 'to the community its true nature, its authentic experience and hidden destiny' (140). National past, present and future are thus united in the inventions of an élite cadre of scholars and artists who develop new methodologies and new symbols designed to serve the notion of an 'authentic identity beneath the alien accretions of the centuries' (75). Thus, although formed on the basis of an authentic, vernacular *culture*, ethno-nationalism nevertheless comprehends a significant *political* dimension, as it was calibrated towards the eventual coincidence of cultural and political discourses.

The civic nation, on the other hand, looked back to a golden age of Graeco-Roman civilisation, and is related to Western European neo-classical concerns with citizenship and order. This form of nationalism comprised discourses of territorialism, participation, citizenship and civic education (116ff.), and it was derived from the professional intelligentsia's desire to connect the language and symbolism of artist-intellectuals with the institutions and activities of a modern state. This model of official state-driven nationalism is akin to the one described by Giddens:

> The nation-state, which exists in a complex of other nation-states, is a set of institutional forms of governance maintaining an administrative monopoly over a territory with demarcated boundaries (borders), its rule being sanctioned by law and direct control of the means of internal and external violence. (1987: 121)

Like Todorov, Smith sees these two forms of nationalism combining and evolving in particular instances throughout the world, although the two remain distinguishable both in theory and in political implication. Ethno-nationalism tended to appeal more to subaltern or submerged communities in that its ruralist, medievalist impulse was available (and thus attractive) to artist-intellectuals from the peripheries of classical Europe. Civic nationalism, on the other hand, evolved amongst the major European ethnic populations into a form of official or territorial nationalism grounded in the bureaucratic state. Both forms, nevertheless, have crucial implications for the modern world, in terms of the ways subjects were imagined to function in communities, and in terms of the ways in which decolonising subjects set about organising discourses of resistance.

Thus theorised, nationalism is a fundamentally unstable discourse, conceived as an attempt (like Romanticism and Enlightenment) to shore up a crisis in European experience during the eighteenth century resulting from challenges to traditional discourses of authority and identity. European nationalism attempted to materialise a relationship between Self and Other, and from the mid-eighteenth century a vast array of cultural and political discourses began to service the idea of the nation-state as the fundamental unit of legitimate international law. As a set of political and cultural prescriptions it is characterised by the drive towards coherency and consistency, yet it cannot help revealing at all times an inability to ground the nation in secure temporal and/or spatial co-ordinates. As a result, generations of nationalists have been constantly embarrassed by the fact that it is capable of being deployed for radically dissimilar ends. The dual nature of nationalism had particularly serious implications when, as in the case of Ireland, it became articulated to a discourse of colonial resistance.

The second factor which has come to engage contemporary theorists of anti-colonial movements is the provenance of the conceptual components deployed in decolonising discourse. Frantz Fanon was one of the first theorists to explore the ways in which nationalism operates in decolonising contexts, and to insist on the implicitly *derivative* nature of nationalist discourses. For Fanon, national independence was a myth as both anti-colonial nationalisms and the postcolonial formations to which they lead remained locked into Western imperialist modes of thought. He discerned a recurring pattern in Third-World nationalism, in which the intellectual and social élites mainly responsible for organising peoples into an effective nationalist resistance were quick to reinstate the systems of hierarchy and privilege which had characterised the colonial polity, only this time with a thin veneer of nationalist respectability (1967: 19–65). They are enabled to do this precisely because the resistance which they organised was conceived in the 'nationalist' terms provided by the colonial power. Official bourgeois nationalism tends to be concerned with a range of specific temporal and spatial discourses which it has adopted uncritically from the colonial power, organised around notions such as tradition, authenticity and sovereignty. This act of betrayal, moreover, straddles both the cultural and political realms, for at the same time as 'the native intellectual is anxiously trying to create a cultural work' and failing 'to realize that he is utilizing techniques and language which are borrowed from the stranger in his country' (180), the national bourgeoisie discover that 'nationalization quite simply means the transfer into native hands of those unfair advantages which are a legacy of the colonial period' (122).

The postcolonial nation thus becomes a 'caricature' (141) of the established European nation-states, and Fanon's contempt for the bad faith of the national bourgeoisie, and the intellectuals who serve it, has set the tone for many similar attacks in recent years. The Indian critic Ashis Nandy, for example, has identified derivativeness as 'the ultimate violence which colonialism does to its victims' and thus the major flaw at the heart of decolonising discourse (1983: 3). Slipping from the public realm of organisation and intention into the realm of psycho-affective function, Nandy claims that colonialism is a particularly insidious form of modern power relations in

> that it creates a culture in which the ruled are constantly tempted to fight their rulers within the psychological limits set by the latter. It is not an accident that the specific variants of the concepts with which many anti-colonial movements in our times have worked have often been the products of the imperial culture itself and, even in opposition, these movements have paid homage to their respective cultural origins. (3)

Even those 'nativist' forms of resistance which are most militantly set against colonialism are culpable in this regard. As Robert Young writes: 'all such arguments, whether from colonizer or colonized, tend to revolve around the terms which the colonizers have constructed. To reverse an opposition of this kind is to remain caught within the very terms that are being disputed' (1990: 168). For many theorists, colonial nationalism is thus a classic 'reverse discourse', fully implicated in the reproduction and survival of that which it disdains.

The question of the existence or extent of colonial nationalism's derivativness has led many contemporary commentators to dismiss such movements as hopelessly incoherent and contradictory. Something of the disdain with which First-World (and indeed many Third-World) intellectuals construe colonial nationalism resonates through Anthony D. Smith's otherwise scrupulously balanced account: 'Truly, colonial nationalisms are still-born; they are imitative "nationalisms of the intelligentsia", unable to forget real nations ... Have not African and Asian intelligentsias imbibed their nationalisms abroad and used them to "invent nations where none existed"?' (108–9).

Liberal and Radical Decolonisation

Both these factors – the dualistic nature of nationalism and the derivative nature of anti-colonial discourse – feed into the modular history of Irish decolonisation that I wish to describe here. With its roots in the same crises of knowledge and identity which gave rise to Enlightenment and Romanticism, nationalism is organised around a fundamental split between discourses of *sameness* and *difference*, and this in turn gives rise to two main forms, strategies or modes of colonial resistance.

The first such mode is one I shall call *liberal*, and will be used to refer to a form of resistance in which subordinate colonial subjects (whether native or settler) seek *equality* with the dominant colonialist identity. In the terminology of modern resistance theory it is characterised by the demand for 'equal access to the symbolic order' by the subordinate subject.[5] As a strategy it is derived from all those 'good' nationalisms described above – Giddens's 'democratic ideals of enlightenment', Todorov's 'ethnic' and 'internal' nationalisms geared towards benign universalism, Smith's medievalist-inspired ethno-nationalism. In the Irish context we shall find that resistance cast in this liberal mode tends to be the province of intellectuals from the settler Anglo-Irish community – such as the 'Patriots' of the late eighteenth century – concerned to 'raise' the status of the colony/nation up to that of the imperial centre, even as they look to maintain the hegemony of their own privileged fraction.

Liberal decolonisation is also 'good' in the sense employed by the discourse analyst Michel Pêcheux in his theory of subject identification through language use, a system which offers a highly suggestive analogy for present purposes:

> The *first modality* consists of a superimposition (a covering) of *the subject of enunciation and the universal subject* such that the subject's 'taking up a position' realises his subjection in the form of the 'freely consented to': this superimposition characterises the discourse of the 'good subject' who spontaneously reflects the Subject ... (1982: 156–7, original emphases)

The resisting subject, that is, *identifies* with the universal subject, thus remaining trapped within the terrain of discourse in which the possibilities of marginality and dominance have been formulated. This is the point at which the other factor bearing on modern decolonising discourse – derivativeness – begins to make itself felt. Liberal decolonising discourse is problematic in that the *equality* to be achieved is already overwritten by the values of the dominant subject, and the language in which equality can be achieved is thus always inscribed with, because formed on the basis of, *difference*. Liberal, egalitarian and universalist strategies, therefore, can be of only limited success because even 'victory' in these terms necessitates the colonial subject's engagement with discursive systems which confirm the original opposition between coloniser and colonised. As Albert Memmi wrote: 'The first ambition of the colonized is to become equal to that splendid model [the colonizer] and to resemble him to the point of disappearing in him' (1974: 120), where 'disappearance' amounts to continued native/settler subservience to colonial domination. For the liberal subject trying to *raise* the experience of the colonised up to that of the coloniser, or to locate a non-ideological realm in which coloniser and colonised can converse in an innocent universal language, *equality* ultimately signifies a denial of national validity and adherence to a structure of differences which maintains the economy of power in favour of the coloniser.

The second mode of decolonisation – which I shall call *radical* – emerges from nationalism's other impulse, towards difference, aggression, anti-universalism and the complete coincidence of cultural and political entities under the aegis of the bureaucratic state. Again, Pêcheux's description is apposite:

> The *second modality* characterises the discourse of the 'bad subject', in which the *subject of enunciation* 'turns against' *the universal subject* by 'taking up a position' which now consists of a *separation* (distantiation, doubt, interrogation, challenge, revolt ...) *with respect to what the 'universal Subject' 'gives him to think'*: a struggle against ideological evidentness on the terrain of that evidentness, an evidentness with a negative sign, reversed in its own terrain ... In short, the subject, a 'bad subject', *counteridentifies* with the discursive formation imposed on him by 'interdiscourse' as external determination of his subjective interiority, which produces the philosophical and political forms of *the discourse-against* (i.e., *counter-discourse*) which constitutes the core of humanism (anti-nature, counter-nature, etc.) in its various theoretical and political forms, reformist *and* ultra-leftist. (1982: 157, original emphases)

As a strategy, decolonisation cast in this radical, 'counter-identification' mode is concerned with what is imagined as unique and different about national identity. Radical decolonisation involves the rejection of imperial discourse, a celebration of difference and otherness, and the attempted reversal of the economy of power which constructs the colonial subject as inferior. In its more militant moments this second mode came to register as a need to shed, violently if need be, the material and intellectual trappings of subordination, and to embrace/construct instead a pristine pre-history which would serve as both Edenic cause and Utopian goal of nationalist activity.

It was a form of this radical decolonising discourse which was adopted by the Irish bureaucratic state in the years after 1922.[6] In as much as this second mode was concerned with the often violent overthrow of English imperial domination, it has been predominantly associated in modern Irish history with the Gaelic – or what was referred to in the cultural debates of the early twentieth century as the *Irish-Irish* – element of the nation. Any Anglo-Irish subject wishing to embrace this politico-cultural option would always find it difficult to gain full access to those discourses and practices from which the radical decolonising gesture emerged. In spite of this, some of the more significant moments in the genealogy of this radical mode that we shall encounter in Chapter 3 are marked by the writings and activities of Anglo-Irishmen such as Thomas Davis and Douglas Hyde.

Again, however, the question of derivativeness is raised. The standard postcolonial line is that at the same time as it affirms the value and validity of *otherness*, radical decolonisation remains 'caught within the very terms that are being disputed' (Young, 1990: 169), implicitly confirming the regime of discourse which constructs the colonial experience as oppositional in the first place. Fanon's description (in *The Wretched of the Earth*) of the radical decolonising intellectual, vainly trying to articulate the nation 'in the light of a borrowed aestheticism and a conception of the world which was discovered under other skies' (1967: 179), and stamping 'techniques and language which are borrowed from the stranger in his country ... with a hallmark which he wishes to be national, but which is strangely reminiscent of exoticism' (180), continues to haunt modern decolonising practices and the postcolonial theory which serves them.

A number of points are worth emphasising at this stage. First of all, liberal and radical decolonisation, obviously opposed in crucial ways, are nevertheless united in their dependence upon a range of specific temporal and spatial co-ordinates which are adopted uncritically from the colonial power, organised around notions such as tradition, authenticity and sovereignty. In ontological, epistemological and ethical terms, both modes emerge from a specific moment of crisis in the history of Western thought. Crucially, both are also versions of, even as they represent opposing engagements with, the autonomous Western subject developed during the eighteenth century. It is in the name of this subject that both modern colonialism and anti-colonialism emerge. Western modes of thought continue, as Pêcheux writes, '*to determine the subject's identification or counter-identification with a discursive formation in which he is supplied with the evidentness of meaning, whether he accepts or rejects it*' (1982: 158, original emphasis). A crucial question follows for any analysis of modern Irish history: how has this crisis, to which Enlightenment offered such an ambiguous resolution, been negotiated in decolonising discourses in a country such as Ireland with such an obscure connection to the mainstream European tradition?

Secondly, these modes (theoretically if not historically distinguishable at this stage) are deemed *necessary* but *insufficient* to the decolonising task: necessary, because the discursive terms made available by the colonial power remain the only means through which a narrative of decolonisation can be articulated (it is in this context that Neil Lazarus (1994: 198) has noted the 'continuing indispensability of national consciousness to the decolonising project'[7]); insufficient, because both before and after independence, the decolonising subject's engagement with received discourses binds her/him to the imperial power against which 'freedom' must constantly be measured. It follows that any discourse of decolonisation which engages with nationalism is bound to be a highly ambivalent practice. On the one hand it may provide opportunities for subjects to articulate their anti-colonialism in a choice of liberal and radical registers. On the other hand, adherence to the modes of thought made available by the colonial power ensures that such anti-colonial discourse is always overwritten with imperialist values.

Following on from these points, a third set of issues is raised, one with which the remainder of this chapter is concerned. Once positioned as subordinate within a discursive economy of power and knowledge, how can individuals and groups strive for release from subjugation without at the same time accepting their designation as Other and thereby reinforcing the structures of that economy? How can the colonised subject articulate difference without metamorphosing into the image of that which is being opposed? How, to adapt the words of Gayatri Chakravorty Spivak, can the subaltern speak?[8]

Decolonisation and Poststructuralism

Along with the critique of liberal and radical modes of anti-colonial discourse, this latter question is the one that most engages modern postcolonial writers, theorists and critics: that is, the possibility of recovering forms of resistance, as well as imagining future politico-cultural formations, which evade, problematise or displace the terms of the encounter between coloniser and colonised. The possibility of another, *third* mode of resistance beyond the limitations of liberalism and radicalism has been the subject of much debate amongst postcolonial intellectuals; indeed, in as much as decolonisation engages with the same crisis of knowledge that structures other dominant forms of modern identity, the question of the limits of dialectical thought is one it shares with many contemporary narratives of resistance (Smith, 1988: 56–69). In each case, the marginal subject finds herself positioned within a framework which limits the possibility of resistance to narratives of *similitude* and *difference*; strategies to go 'beyond' or 'outside', or posit a discursive realm 'before', the law of oppositional logic appear merely to reaffirm the framework and the economy of power/knowledge on which it relies. Cast in these terms, there is in fact no 'outside' or 'beyond' to which the colonial subject can appeal or escape. Although presence, essence and identity may be said to be at the root of the colonial problem, they remain the only available tools for decolonising activity.[9] The question insists: how can the subaltern subject and community deal with received oppositions given that there is no language they could speak, no activity in which they could engage, which could not somehow be narrativised and recouped by this oppositional cast of mind?

Fanon sought hope in the possibility of a new, post-European humanism, claiming (in *The Wretched of the Earth*) that 'After the conflict, there is not only the disappearance of colonialism but also the disappearance of the colonized man. This new humanity cannot do otherwise than define a new humanism both for itself and for others' (1967: 198). This reiterated a point already made in *Black Skin, White Masks*:

> The Negro is not. Any more than the white man.
>
> Both must turn their backs on the inhuman voices which were those of their respective ancestors in order that authentic communication be possible. Before it can adopt a positive voice, freedom requires an effort at disalienation ... It is through the effort to recapture the self and to scrutinize the self, it is through the lasting tension of their freedom that men will be able to create the ideal conditions of existence for a human world. (1986: 231)

The conventionalism and apparent sentimentality of Fanon's discourse in these instances seems difficult to reconcile with some of his more hard-line statements elsewhere. This particular aspect of his thought has been under revision for a number of years, mostly from theorists whose own critical visions are coloured by broadly poststructuralist systems of thought. For many such critics, the most interesting and promising approach to the problem of decolonisation is predicated not on a rejection of, or alternative to, the identitarian discourse of liberal and radical modes, but on methodologies which reveal their essentially *displaced* and *performative* nature. Homi Bhabha, for example, has engaged throughout his career with the dualistic model described above, criticising liberal and radical decolonising initiatives even as he attempts to explain their political and psychological dynamics. But despite genuine admiration for Fanon's political and intellectual example, Bhabha has little time for what he describes as the former's 'banal existentialist humanism', or his 'Hegelian dream of a human reality *in-itself-for-itself*' (1986: xx and xxi). Bhabha's own highly influential researches have been based upon a complex (often obscure) enmeshing of high European theory and subversive subaltern politics.

Bhabha formulates the split between liberal and radical decolonisation (as it is registered in the realm of critical-aesthetic discourse) as a pseudo-choice between 'universalism' and 'nationalism' (1984: 99ff.). Like Fanon, he understands these modes to be both natural and necessary responses to imperialist discourse. Yet they are also ultimately inappropriate for any discourse of decolonisation. This is because the stand-off between universalism/liberalism and nationalism/radicalism is fought essentially on the same aesthetic ground – that of 'image analysis'. In both cases, 'The "image" must be measured against the "essential" or "original" in order to establish its degree of *representativeness*, the correctness of the image. The text is not seen as *productive* of meaning but essentially reflective or expressive' (100, original emphases). 'Image analysis' is thus part and parcel of an Enlightenment-sponsored 'hermeneutics of suspicion', motivated by the promise of a 'truth' to be discovered beneath the representation. Such 'truths' are always relative to the world-view of the dominant formation, however, so that the independence struggles of decolonising formations invariably take place 'within structures partly inherited from colonisation' (*Passages*, 1994: 41). Mortgaged to image analysis and the hermeneutics of suspicion, liberal and radical cultural-critical strategies are limited in their effectiveness because they always acknowledge, however remotely or tacitly, the discursive limits preset by the colonising power.

Against both image analysis and Fanon's humanism Bhabha attempts to uncover 'a mode of negation that seeks not to unveil the fullness of Man but to manipulate his representation. It is a form of power that is exercised at the very limits of identity and authority, in the mocking spirit of mask and image' (1986: xxiii). Representation segues irresistibly from cultural-critical analysis to political prescription, and thus the basis is laid for another mode of decolonising discourse, one that we might name (although Bhabha himself uses the word sparingly) *hybridity*. Bhabha's claim is that colonial cultural discourse has been produced in

> a 'separate' space, a space of *separation* – less than one and double – which has been systematically denied by both colonialists and nationalists

who have sought authority in the authenticity of 'origins' ... as discrimination turns into the assertion of the hybrid, the insignia of authority becomes a mask, a mockery. (1994: 120, original emphasis)[10]

Bhabha has insisted that hybridity is an effect of the history of colonialism and the myriad encounters between differently empowered subjects; but also, and more importantly, it is the condition of language itself, the 'Third Space' (1994: 37) that marks the moment between reality and representation, between the *performance* of the cultural text and the reality to which it refers (19–39). Such a strategy of resistance constitutes a dialectical movement between a radically ambivalent colonial discourse (always already doubled – self-fashioning *and* self-erasing) and a similarly ambivalent decolonising discourse which is also always attempting to mask its ineradicable derivativeness. In other words, colonial resistance is hybridised both *before* and *because of* the colonial encounter. It is the enabling, if confusing, possibilities ensuing from the movement between these two moments with which Bhabha is concerned in his work:

For a willingness to descend into that alien territory ... may reveal that the theoretical recognition of the split-space of enunciation may open the way to conceptualising an *inter*national culture, based not on the exoticism or multiculturalism of the *diversity* of cultures, but on the inscription and articulation of culture's *hybridity* ... And by exploring this hybridity, this 'Third Space' we may elude the politics of polarity and emerge as the others of our selves. (1994: 38–9, original emphases)

What this Third Space provides is the possibility of a politics of *displacement*, an operation whereby the given categories are (necessarily) performed, but in such a way as to question their givenness, authenticity and originality.[11] It is the very character of decolonising discourse – doubleness and derivativeness – which enables the deconstructive act; this act, in turn, reclaims the dissident voices of the past and the subversive potential of the present. This Third Space, moreover, emerges during every colonial encounter, but is either repressed or recast into narratives of exoticism, danger or fantasy by the dominant formation. It is a space in which the coherent, transparent subject developed by Western discourse comes under pressure in that moment of differentiation between identity and otherness on which colonial discourse turns. Decolonising critics must not only know where and how to look for this disruptive scene; once located, it must be read differently, against the grain of a hermeneutics which in its interpretative assumptions – authority, intention, identity – always reproduces the 'reality' of colonialist discourse.

Conceived thus, hybridity encompasses a history, a politics, an aesthetics and a critical theory of decolonisation. Strategically deployed, it reveals both the doubleness at the heart of Western culture *and* the collusive nature of reverse discourses such as Western-inspired colonial nationalism. Theory has opened up a space between signifier and signified, and it is precisely this Third Space – 'unhomely' (10), 'unthought' (64), 'ambivalent' (92), 'uncanny' (101), 'undecidable' (136) – which allows for the emergence of effective resistance.[12] Western thought aims for originality and presence but is undone by its necessary repetitive and translational status, caught always and everywhere in the undecidable shift between imitation and identification: 'It is at this moment of intellectual and psychic "uncertainty" that representation can no longer guarantee the authority of culture; and culture can no longer guarantee to author its "human" subjects as the signs of humanness' (137).

Gayatri Chakravorty Spivak is another theorist and critic who has been concerned to test the limits of anti-colonialism. Spivak has been especially instrumental in mapping the disjunction that appears between state-building national élites and certain modes of resistance which, as described by Benita Parry, are not 'calculated to achieve predetermined political ends or to advance the cause of nation-building' (1994: 173). At the same time, she disavows any nostalgia for the subaltern subject or the intellectualist project of recovering marginalised voices from the past which might somehow speak the language of an authentic native resistance. Instead, Spivak has advanced the possibility of a 'strategic essentialism' with regard to the *données* of the Western tradition, a system of imaginative negotiations in which the decolonising subject attempts to change from within a situation they are 'obliged to inhabit' (1990: 72). These subjects work with systems and tools whose collusion they comprehend but which nevertheless constitute their only means of engagement; they operate with a form of knowledge *belonging* to the dominant discourse but necessarily changed through its articulation to subaltern contexts and idioms.

Spivak finds the exemplary form of such 'strategic' knowledge in that practice which, she claims, has always spoken Western culture's doubleness: literature. 'Textuality', she claims, marks 'the unavailability of a unified solution' (1987: 78). As Robert Young explains, it is the originary scission encoded into Western history and exemplified in Western textuality that 'enables the deconstructive move which displaces as well as reverses an opposition, such as that between colonizer and colonized, and thus provides a position for the forcing of an effective critical leverage (1990: 169).'[13] The force of Spivak's position, as Young continues, is that effective colonial resistance requires not alternative or counter-knowledges, but strategies which 'contest and inflect the more far-reaching implications of the system of which they form a part' (172). Moreover, because criticism and theory are themselves integral to that system, Spivak's own discourse has to remain deliberately under-formulated, self-conscious and 'supplementary' of itself if it is successfully to expose the myths (such as presence, autonomy and representation) upon with the project of Western rationalism is founded.

These modes of decolonisation explored by Bhabha and Spivak represent, then, not so much an alternative to liberal and radical discourses, but rather a staging of the rules, codes and languages which provide their conceptual coherence. These strategies of resistance are still possible, indeed, they are necessary stages within the narrative of decolonisation. However, the final trope in that narrative requires a performance of all previous decolonising discourse such that the colonial subject can experience identity as both pressing reality and staged event, as both the basis for decolonising activity and the fiction, the rhetoric, the myth which enables that activity. Only in this way can the colonial subject 'break out of' the disabling bind of doubleness and derivativeness. As the quotation marks indicate, however, this 'breaking out of' will also entail a 'breaking into', with the colonial categories maintained as reference points even as their constructed nature is revealed.

Both these positions, given their bases in what we might call 'textualism', have implications for the theory of anti-colonialist criticism I shall be exploring in the next chapter. Whether Bhabha succeeds in redeeming 'the pathos of cultural confusion into a strategy of political subversion' (1986: xxii), or whether Spivak really demonstrates, as Young claims, 'the possibility of providing a critique in which both theory and detailed historical material can be inflected towards an inversion of the dominant structures of knowledge and power without simply reproducing them' (1990: 173), remains unclear.[14] Having noted that the typical methodologies deployed by Bhabha and Spivak, as well as the overall deconstructive ethos of their work, have their bases in a broadly defined poststructuralism, especially the work of Jacques Derrida,[15] it is clear that the relationship between postcolonial and post-structuralist theories remains politically uncertain, to say the least.

As certain versions of the latter have come under attack, so a number of objections to both the form and the content of post-structuralist-inspired analyses of anti-colonialism have emerged since the 1980s.

One of the earliest notes of discord was sounded by the figure most instrumental in establishing colonial discourse as an important area of study in its own right, Edward Said. Whereas the ground-breaking *Orientalism* was clearly poststructuralist (mid Foucauldian) in methodology and ethos, Said has of late been drawn back towards the more humanistic stance of Fanon, finding himself at odds with the 'hopelessly tiresome' textualism which characterises many versions of contemporary postcolonial theory.[16] At the end of *Culture and Imperialism*, Said attempted to map out his own 'Third Space', but one that was much more answerable to the concerns of traditional scholarship and a more liberal model of the ways in which culture feeds into politics. Against both collusive universalism and those independence movements in denial about the impact of imperial discourse on national history, a properly 'liberationist' discourse should proceed along a specific intellectual-methodological trajectory:

> First, by a new integrative or contrapuntal orientation in history that sees Western and non-Western experiences as belonging together because connected by imperialism. Second, by an imaginative, even Utopian vision which reconceived emancipatory (as opposed to confining) theory and performance. Third, by an investment neither in new authorities, doctrines, and encoded orthodoxies, nor in established institutions and causes, but in a particular sort of nomadic, migratory, and anti-narrative energy. (1993: 337)

Said is in fact one of the principal figures (another is Julia Kristeva) offered by Christopher Norris as examples of the back-tracking undertaken by some of the more conscientious of modern theorists in the face of poststructuralism's apparent elision of history and ethics in favour of a 'textualism' which often amounts to little more that 'sophistical chicanery' (1994: 115).[17] By reducing history to discourse, and truth to a rhetorical position adopted within the text, 'post-structuralism has removed the very possibility of reasoned, reflective, and principled ethical choice' (109). This results from a misunderstanding (and subsequent misuse) of the truly radical implications of Derrida's thought, as well as contemporary theorists' ignorance of their own philosophical pre-history.[18] Said's insight (one which according to Norris also struck the later Foucault) has been to recognise that the 'distinction between historical fact and literary or fictive representation has been vital to the entire post-Renaissance enterprise of enlightened secular critique' (112), and that critical initiatives which deny this distinction, having cynically abandoned reason and progress as mere rhetorical ruses, are incapable of articulating 'an ethics and a politics possessed of genuine emancipatory values' (125).

Even the ethico-political spin put upon poststructuralism by theorists such as Bhabha and Spivak – to the effect that textualist strategies have critical implications for the practical operation of colonial power relations – is dismissed by Norris as so much intellectualist attitudinising:

> Nor are the current alternatives very much better when, as often happens, they start out from the same *a priori* persuasion – that is to say, from some variant of the linguistic or textualist turn – and seek to pass 'beyond' it while leaving its major premises firmly in place ... If history can be transformed into yet more grist for the textualist mill then the same applies to those ethical modulations ... which have lately been invoked as a high-toned alibi for theorists in quest of some opening beyond the post-structuralist prison-house of discourse. Such an opening is simply not available ... so long as they persist in the dogma that subjectivity, knowledge, and experience are relative to (or 'constructed in') language, and hence that genuine respect for the other must entail casting doubt upon everything that enables us to respond or communicate across or between cultures. (119–20)

Norris feels that it would be a mistake, therefore, for critics from ex-colonial nations to become ensnared in the sophistries of post-structuralism or its various disciplinary offshoots: 'Indeed there is a sense in which the colonization of historiography by literary theory – or by "radical" ideas imported from that field – can serve to obscure both the historical realities of colonial oppression and the experience of those who either suffered its effects or protested its injustice' (109). The task of these postcolonial critics, rather, is to return to history and to the Enlightenment as the basis for ethical, reasoned, and most importantly *effective* critique.

Such an argument is supported by those favouring a return to ethical historiography as the basis for an analysis of imperialism and decolonisation. There would appear to be a growing feeling that the 'Hostility to nationalism exemplified by certain critics and theorists comes into focus as a kind of radical elitism' (Lazarus, 1994: 214), and that poststructuralism contributes to a situation in which the baby of effective resistance is in danger of being thrown out with the bathwater of bourgeois ideology. In the course of producing his own tripartite model of nationalism, for example,

Partha Chatterjee has traced the ambiguities and contradictions attending non-European nationalism as a result of the interaction between received and indigenous discourses of resistance.[19] Chatterjee has elaborated the established notion of nationalism as a derivative discourse, a claim upon which, as we have seen, so much modern postcolonial theory turns. Such a notion, he claims, denies coherence, agency and effectivity to the colonial subject. While it seems clear that nationalism was derivative of the formation of discourses which constituted the colonised as oppressed subjects, and while the postcolonial drive towards statehood obviously leaves it exposed to new, global forms of subjectification, it also seems clear that these discourses were necessarily altered when articulated to various forms of decolonising politics.[20]

By the same token, acknowledging the validity of native identity does not amount to an unqualified celebration of the national essence or a disparagement of extra-national effects as manifested in the cultural history of the nation. Attacks on nationalism couched in the tones of poststructuralist critiques of identitarianism have by and large failed to register the struggle *within* nationalism itself with regard to the limits of the nation and the politico-cultural policies attending the discourse of decolonisation. Neither can the poststructuralist critic object to the unity of the postcolonial nation-state, as Derrida has said, 'simply because it is the result of a process of unification' (*Passages*, 1994: 41), or because such formations are (contrary to their self-images) demonstrably entwined with Western imperialism. Poststructuralism's fix on *alterity* may in fact be seen to have less in common with postcolonialism than with the third element of the great modern 'postal' triumvirate: postmodernism. The decentred subjects and nonmimetic narrative modes of colonial history, as Kumkum Sangari maintains, 'inhabit a social and conceptual space in which the problems of ascertaining meaning assume a political dimension qualitatively different from the current postmodern skepticism about meaning in Europe and America'. In such a reading, poststructuralism actively 'contains' all modes of alienation and all anti-essentialist discourses in terms of a specific Western crisis, thus encouraging 'a naive identification of all nonlinear forms with those of the decentered postmodern subject' (1995: 143, 146).

All told, the nexus of postcolonialism and poststructuralism is, at best, troubled, and many critics have as a consequence become engaged in the search for alternative routes out of the theoretical and practical crises which beset decolonisation. Nowhere more so than in Ireland, a country possessing such an uncertain relationship with both European history and contemporary cultural-critical theory.

Decolonisation in Ireland

Versions of the debates and models described above have emerged in contemporary Irish cultural criticism. Because of the peculiar circumstances of Irish history, such models are inflected in particular ways, some of which I shall be identifying and elaborating upon during the course of this study. But to finish this chapter, I want to mention briefly the means by which three such models attempt to outmanoeuve the limitations of established resistance discourses.

In his edited collection on 'the Irish mind', the philosopher Richard Kearney places the crisis in eighteenth-century European thought as an effect of the orthodox dualist logic of *either/or* encoded in Western philosophy since ancient times. This logic is based on three principles: '(i) A is A (the principle of Identity); (ii) A is either A or non-A (the principle of the Excluded Middle); (iii) if A is A it cannot be non-A (the principle of Contradiction)' (1985: 296). Kearney then suggests that 'the Irish mind' may be seen to favour a different logic, one organised around the principle of *both/and*, and characterised by an intellectual ability to hold the traditional oppositions of classical reason together in creative confluence. For Kearney, the 'Irish mind' may be traced throughout the history of Irish writing in English – Swift, Sterne, Berkeley, Wilde, Shaw, Beckett, O'Brien, Behan, and so on, with James Joyce nominated as the seminal modern exponent of this doubled, disruptive discourse.

Kearney's model is primarily culturalist in orientation, based on the conviction that political decolonisation must be accompanied by a 'decolonisation of the mind' if it is to have any lasting significance. The categories of colonial subjectivity, such as Irishness and Englishness, cannot be replaced, because they represent the only terms in which a decolonising discourse can be articulated. It may be, to adapt Fanon, that 'Irishness is not'; but if 'Irishness' is refused as an imperial invention (or a nationalist refinement of that invention) then there are no possibilities for engagement and change. Kearney's point is that even as Joyce rehearses the traditional spaces and practices wherein an Irish identity operates, he indefinitely postpones a final, finished decolonising narrative peopled by authentic Irish subjects. In this reading, Joyce's work is simultaneously constructive and *de*constructive, radically aware of language's role as primary site and emblem of the decentred subject, yet enabled through his access to an Irish cultural consciousness ('the Irish mind') to appreciate the supreme materiality of language and the radical positionality of the truths it speaks. The 'Irish mind' is thus simultaneously *engaged* – in decolonising activity of various (liberal and radical) kinds – and *displaced*, self-reflexive, always aware of its own arbitrary location in time and space. For Kearney, therefore, Irish culture operates in terms of a rhythm of affirmation and scepticism, and it becomes possible for the critic to identify and trace instances of this rhythm throughout Irish history.

Such a model brings its own difficulties, however. Kearney cannot escape the possibility that this dialectical logic (which he *opposes* to classical dualistic logic) could itself become the *sign* of Irish otherness, thus once again becoming subsumed into the oppositional, identitarian politics of imperialism and nationalism. The idea of an 'Irish mind' opposed to the constant banality of *either/or* scenarios might be thought by many to be a peculiarly consoling notion, an exotic realm of thought and activity into which the exhausted Western imagination can occasionally slip before returning, refreshed and reassured, to the rigours of 'real' dialectical thought. The Irish mind is thus in danger of becoming permanently 'hybridised', identified precisely by its inability to interact with established, 'normal' reality. Joyce becomes the archetypal Irish writer precisely to the extent that his aesthetic practice is removed from reality. Seamus Deane puts it this way:

> A literature predicated on an abstract idea of essence ... will inevitably degenerate into whimsy and provincialism. Even when the literature itself avoids this limitation, the commentary on it re-imposes the limitations again ... The point is not simply that the Irish are different. It is that they are absurdly different because of the disabling, if fascinating, separation between their notion of reality and that of everybody else. (1985: 57)

It may thus be seen that instead of contributing to a re-distribution of power relations, the notion of an Irish mind characterised by an alternative *both/and* logic could instead feed the stereotypical assumptions which helped to create asymmetrical power relations in the first place. The moment in which established reality is questioned is always in danger of hardening into a strategy, a badge of otherness, the very 'sign' of difference. Just as every tactic employed by the patient to refuse the analysis can be explained by the analyst, so the refusal by colonised subjects such as Joyce to limit themselves to the reality provided by the colonising power can eventually be diagnosed as a typical colonial response, another brick in the wall forming the border between imperial self and colonial other.

Kearney's work emerges from, and engages with, an established European philosophical tradition and can thus be accommodated within established intellectual discourses, Irish or otherwise. Although a number of contemporary Irish cultural critics have attempted to engage with recent developments in postcolonial theory, they have found themselves confronted with a critical establishment entrenched in 'image analysis', or what David Lloyd has called the 'narrative of representation'.[21] Lloyd himself considers this to be unfortunate, not only for what such theories can bring to Irish studies by way of a revitalised intellectual and methodological programme, but also for what Ireland's unique colonial experience can bring to a frequently complacent and generalising postcolonial theory:

> Any serious analysis of Ireland's complex relation to colonialism must draw on the international histories and analyses of colonial processes and ideologies, not in order to throw up facile analogies but in order to comprehend more deeply the differentiated processes of domination and the insistence of alternative structures of cultural practice. (1997: 91)

Lloyd's own work engages with the modular theories explored throughout this chapter, concerned like many postcolonial theorists to analyse the relations between the official nationalism promulgated by state-building élites, and modes of resistance which eluded attempts at official narrativisation. Yet, while on the one hand criticising the 'inadequacy of the model of modernisation that structures western pronouncements on nationalism, development and culture' (91), on the other he has been instrumental in advancing the standard 'derivativeness' line of postcolonial theory with regard to Irish cultural history, pointing out that 'while nationalism is a progressive and even a necessary political movement at one stage of its history, it tends at a later stage to become entirely reactionary, both by virtue of its obsession with a deliberately exclusive concept of racial identity and, more importantly, by virtue of its formal identity with imperial ideology' (1987: x). In his insistence that the identitarian discourses of dominant nationalism are not the solution to colonial violence but the precise location of the problem, Lloyd reveals the poststructuralist assumptions underpinning his own version of postcolonial theory. At the same time, he has been concerned both to undertake and encourage research which is 'engaged by the fine grain of the alternative narratives and practices embedded in Irish cultural history' (1997: 91), and to trace modes of refusal which problematise, even as they engage with, the dominant narratives of decolonisation. Typically of postcolonial theory, Lloyd sees effective resistance to imperial domination residing more in fragmentary and hybridised discourses than in fully rational politico-cultural initiatives which are constituted in response to (and therefore in collusion with) the oppositional logic of imperialism.

This two-pronged programme for a revitalised Irish Studies operates throughout the essays collected in *Anomalous States: Irish Writing and the Post-Colonial Moment*. The first (negative) aspect can be seen in the notorious demolition of the 'minor' poet Seamus Heaney, whose work is considered 'profoundly symptomatic of the continuing meshing of Irish cultural nationalism with the imperial ideology which frames it' (1993: 37). For Lloyd, Heaney is merely the latest (and far from the most accomplished) in a long line of Irish figures who have come unstuck when confronted with 'the logic of identity that at every level structures and maintains the post-colonial moment' (56).

Against both Heaney's collusive, essentialist fantasies *and* the kind of critical discourse in which these are celebrated, Lloyd is concerned to uncover cultural effects and practices which approach 'the threshold of another possible language with which a post-colonial subjectivity might begin to find articulation' (56). This is the second (positive) aspect of his programme, engaging with those versions of postcolonialism which have looked to evade the *impasse* of received binary thought. Traces of such a language can be found in 'minor literature', in the characteristically irregular and refractory nature of popular culture, as well as in the work of certain modernist writers such as Beckett, Yeats and Joyce. The latter, again, emerges as an especially important figure, and it is around Joyce's work that Lloyd builds a theory of a potentially non-collusive colonial resistance. 'Adulteration' (compared and linked with Bhabha's version of 'hybridity' in a long footnote) helps to reveal not only nationalism's 'mimicry of imperial forms' (123), but also to recuperate those modes of resistance within Irish cultural history which (to quote Parry again) were neither 'calculated to achieve predetermined political ends or to advance the cause of nation-building' nor 'readily accommodated in the anticolonialist discourses written by the elites of the nationalist and liberation movements'. Lloyd describes the principle this way:

> The processes of hybridization or adulteration in the Irish street ballads or in *Ulysses* are at every level recalcitrant to the aesthetic politics of

> nationalism and ... to those of imperialism. Hybridization or adulteration resist identification both in the sense that they cannot be subordinated to a narrative of representation and in the sense that they play out the unevenness of knowledge which, against assimilation, foregrounds the political and cultural positioning of the audience or reader. (114)

Thus considered, adulteration operates as an ambivalent, trouble-making discourse, evading colonialist discourse as well as the dominant forms of nationalism with its fantasies of presence and originality. Adulteration, crucially, is not a critical invention *avant la lettre* but an acknowledgement – against criticism and its focus upon intention and revealed meaning – of the strategies and initiatives of those who have throughout history found themselves on the margins of power. Joyce's achievement, in this reading, was to textualise the incoherencies of Irish history, writing the nation's epic but organising it around the profoundly anti-epic effects of contingency and ambivalence.

Lloyd's concept of 'adulteration' operates in a similar way to the figure of 'allegory' advanced by Luke Gibbons. Working with the standard dualistic modular system described above, Gibbons differentiates between 'European conceptions of nationalism' in which 'a premium is placed on coherence and abstraction, and the clarity of political consciousness comes to resemble the unmediated self-presence of the individual subject' (1996: 137), and this model's 'more idiosyncratic peripheral variants' (135) in which national identity is a 'fugitive' (145) experience constructed by marginalised subjects from the fragments of popular culture. According to Gibbons, Ireland during the nineteenth century witnessed a struggle between these two modes, the one constitutionalist and state-oriented (represented by figures and institutions such as O'Connell, Young Ireland, Home Rule and Parnell), the other a 'dissident, insurrectionary tradition' (146), made up of half-digested epic imagery, the rituals of agrarian secret societies and the performative aesthetics of popular ballads. Although engaging with (and eventually subsumed by) the concerns of dominant (liberal and radical) modes of decolonisation, this latter form of resistance, as Neil Lazarus writes in an analogous context, was 'often entirely divorced from and unassimilable to the "vertical" political concerns of elite anticolonial nationalists' (1994: 207).

These two modes, moreover, also possessed corresponding aesthetic forms. Constitutional nationalism, mimicking the precepts of mainstream European nationalism, depended for its existence on a kind of abstract knowledge derived from print culture, thus finding its characteristic voice in the modern newspaper's collapse of all the nation's minor, coterminous narratives into a single, national metanarrative.[22] This is the narrative – with fully self-present national subjects and highly charged national symbols – that Irish nationalism would tell itself throughout the postcolonial era to ward off that sense of doubleness inherited from the mainstream European tradition. It is in this way that modern anti-colonialists 'mimic their masters' voices, and reproduce in their own idioms the closed, univocal expressions of identity articulated in the imperial centre' (Gibbons, 1996: 7).

However, 'threaded through these totalizing images is a much more complex set of narratives, often figuring the self-image of a culture in allegorical terms, with all the contestation of identity and openness towards the other which that entails' (7). Unlike the unity of culture pursued by constitutional nationalism, that is, allegory articulates 'a fugitive existence in the margins between the personal and the political' (145). Recalling both Bhabha's 'Third Space' and Spivak's 'strategic essentialism', allegory articulates an 'identity without a centre' (134), an identity engaging with, even as it looks to articulate a realm of experience and activity beyond, received possibilities:

> Allegory in an Irish context belongs to the politics of 'the unverbalized'. It is not just a poetic device, but a figural practice that infiltrates everyday experience, giving rise to an aesthetics of the actual ... For allegory to retain its critical valency, it is vital that there is an instability of reference and contestation of meaning to the point where it may not be at all clear where the figural ends, and where the literal begins. (20)

Despite the victory for official state nationalism in Ireland in the early decades of this century, the allegorical tradition provided the 'proto-modernist' (6) strategies from which future Irish artists like Joyce and Beckett learned, thus anticipating the collapse of the subject and of narrative which would herald the cosmopolitan European modernisms of the twentieth century. At the same time, allegory succeeded in insinuating its own dissident effects into the mainstream, and it is possible for contemporary critics to trace these effects in official nationalist discourse. Allegory may be a characteristic postcolonial form but Gibbons is not implying, contrary to some critiques, that its uses and effects are confined to subaltern discourse; rather, like the symbolist and metaphorical figures to which it is formally opposed, allegory ranges over a wide range of cultural and political fields, introducing contradictory inflections into discourses (such as official nationalism) which aim for the traditional narrative effects of unity, identity and closure.[23]

Allegory thus informs certain modes of decolonising activity from the past as well as certain critical activities of the present, for, like Lloyd, Gibbons is keen to advance a model of Irish Studies fully attendant to the unique ways in which culture has engaged with politics in modern Irish history, as well as in colonial and post-colonial formations generally. And for such a task, he suggests, traditional cultural paradigms and methodologies are both inadequate and inappropriate. Analyses unproblematically transposed from Western models of tradition and modernisation to Irish history fail to engage with what is specific about decolonising formations – namely, both the lack of secure traditions and the politically charged nature of discourses of modernity. As Lloyd writes: 'The very division between politics and culture that is the hallmark of liberal ideology is conceptually bankrupt throughout the post-colonial world' (1997: 87). Rather, the task for critics from a country with a fractured, colonial past such as Ireland is to pursue 'the oblique and often recondite ways in which social forces and historical events are inflected by cultural forms, their characteristic figures and narrative patterns' and thus 'give a belated hearing to voices or patterns of experience that have escaped the nets of official knowledge, or have been muted by the dominant ideologies of the day' (Gibbons, 1996: 16).

Kearney, Lloyd and Gibbons, therefore, all describe modular histories of Irish decolonisation. Identifying in each case the *impasse* of liberal and radical modes of resistance, each explores the possibilities afforded by concepts such as 'the Irish mind', 'adulteration' and ' allegory' to discover a mode of decolonisation which might evade or postpone those characteristic 'pitfalls of national consciousness'. Each seems more attuned to Said's late-humanism than to the ironic scepticism and ego-textualism of poststructuralist modes of inquiry; each seems keen to produce counter-narratives which would both *reveal* (in their subject matter) and *exemplify* (in their characteristic methodologies) the typical colonialist experience of difference/derivativeness. In Part 2 of this book I shall be going on to look at a range of examples of each of these modes of anti-colonial resistance. At this point, however, it is necessary to think about the relations between culture and politics in modern Irish history, and more precisely to think about the discourse in which such a relationship has been conceived and re-presented: criticism.

2 Culture, Criticism and Decolonisation

The alert reader will have noticed a steady slippage in the preceding chapter, from 'politics' (describing nationalism as a form of modern political identity predicated on specific relations of power) to 'culture' (foregrounding the role played by cultural discourse in the construction and mediation of these relations). This reflects the wider situation in postcolonial studies in which, following Said's example in *Orientalism*, a great deal of research time and energy continues to be devoted to analysing the productive relations between what he called 'political power in the raw' (1985: 12) and cultural discourses.[1] 'Culture' in this case is taken in the widest possible sense, describing, as Raymond Williams put it, 'a particular way of life, which expresses certain meanings and values not only in art and learning but also in institutions and ordinary behaviour' (1984: 57). We encountered Irish variations on this methodological-conceptual line in Chapter 1 when noting the concern of Kearney, Lloyd and Gibbons to trace the articulation of modern Irish political possibilities through a wide range of cultural activities. And although the dynamics of the relationship between cultural and political discourses continues to be revised and refined, the force of the original point insists across a wide spectrum of contemporary postcolonial debate: culture played a vital role in the colonising and decolonising processes, and certain kinds of (cultural) subject have emerged throughout history to promulgate the various modes of decolonisation. Culture *represents* (in the dual sense of *embodying* and *standing for*) the most salient discourse wherein the battle for power in colonial society occurs; it enables, justifies, and gives form to the hegemonic struggles in which the (colonising and decolonising) subject acts. Cultural discourse, to adapt Foucault, is the power which is to be seized, and a struggle for culture is perhaps the one factor common to all the disparate histories of decolonisation.[2]

But what kind of culture? One problem with such an emphasis in the Irish context is that despite the promise to reconfigure the relationship between culture and politics in colonial history, it represents an essential continuity with older, more established and conservative models. For many commentators it is clear that Ireland has never had to experience the angst of critics from other countries confronted with the nature of the connection between culture and politics. The great theoretical questions which have animated European intellectual debate for the past thirty years regarding the role of culture in the production and reproduction of political effects simply have no purchase in a country where, as an entire academic industry has assured us, culture and politics are to all intents and purposes one and the same thing. Especially after the cultural revival of the 1890s, culture and politics were assumed to be intimately associated realms, and this relationship came to constitute one of the island's dominant self-formative images.

Rather than radicalising the discourse, however, it seems clear that the culture–politics nexus in Ireland has been systematically channelled off into disabling pseudo-debates between established political and cultural positions, and has helped to shore up a fundamentally quietistic model of the relations between the two spheres. Malcolm Brown, for example, states baldly that 'Modern Ireland provides us with the classic case of an impressive literature brought to birth by politics' (1972: vii). The schema is familiar: politics first, literature second. Literature must be read in terms of its engagement with, and representation of, the 'real' empirical identities and discourses established by the 'hard' disciplines of politics, history, sociology. Lest even this seem too radical, however, Brown goes on to qualify his thesis by explaining that

> while our historical reconstruction is going forward, poets must be held mostly under constraint. To take one's Irish history from an uncritical reading of Yeats, Joyce, O'Casey and O'Faolain [sic] is, though convenient, a reckless procedure in any case; and as a base for comparative inquiry ... it is altogether pointless. But to assemble the history elsewhere and then to set it against the poetic version is a productive and enlightening exercise. (ix)

The book's title refers not to something inherent in the literary text, then ('the politics *of* Irish literature'), but to the contexts in which texts are produced and consumed. 'Politics' feeds into the text from the real world of Young Ireland, Home Rule, Parnell, the Land League, etc., where it is mediated by authors employing the specificities of literary language and form to produce a 'version' of the real. Such faith in the divisibility of the political and the poetical has been under pressure at least since the avant-garde artists of the 1920s announced the 'revolution of the word', a manifesto which, as Colin MacCabe writes, 'appropriated the political for the aesthetic and located historic change in the choice of language made by individual writers' (1979: 1). Brown's model of 'the politics of Irish literature' is in fact untenable in the late twentieth century with the acceptance of the politicisation of discourse across a wide spectrum of theoretical debate. Yet, in many respects it is clear that this is the model of political literature and literary politics which by and large continues to hold sway in Irish Studies.

As another manoeuvre towards the same end, consider F.S.L. Lyons' verdict on 'the false assumption' made by some Irish intellectuals at the beginning of the twentieth century ' ... that in art, as in society, collaboration between classes, religions and races would fill the political vacuum. But in reality, there was no vacuum. The political issue – the issue of separation from Britain – remained the central issue and everything else would continue to be judged according to whether it added to or subtracted from the national demand' (1973: 246). In some respects this is a 'radical' position which refuses what David Lloyd calls (as noted in Chapter 1) 'the very division between politics and culture that is the hallmark of liberal ideology' (1997: 87). Such divisions, according to postcolonial (and most oppositional) theory, obtain in formations lacking the historical experience of colonialism, constituting the norm against which a 'radical' culture/politics nexus is asserted. Lyons also appears to refuse the sequestration of culture from politics that characterises 'liberal ideology'; yet he did so from within an Irish society which had been moving steadily towards a traditional Western model of 'liberal' nation-statehood during the twentieth century. In fact, the framework within which he operated was thoroughly bourgeois-liberal in conception, while the terms with which he worked were the established Western ones in which the 'cultural' and the 'political' signified specific, discrete spheres of activity.

If this contradiction has allowed generations of conservative critics to range over Irish literature declaiming its 'political' basis even as they work within systems and with forms which are geared towards the constitution of 'the very division between politics and culture that is the hallmark of liberal ideology', it is also a problem which confronts more self-consciously 'radical' interventions. In the same review of Terry Eagleton's *Heathcliff and the Great Hunger*, Lloyd attacks what he sees as 'The all too familiar habit of regarding the colonized as cultural *producers* whose work must be theorised elsewhere', and also (paraphrasing Chatterjee) 'the inadequacy of any model of modernisation that structures western pronouncements on nationalism, development and culture' (1997: 90–1). Given its peculiar history and its limited cultural resources, Ireland's literary heritage is widely accepted as a wonder of the modern world and a source of constant fascination for international critics. Traditionally, however, it seems that this heritage can only be engaged (as with Eagleton) in terms of 'larger' critical systems and temporalities emerging from the experiences of that 'modern world'. This is ironic because much of the time it is in fact the very nature of the relationship between received metropolitan systems and indigenous initiatives that is exactly the issue in modern Irish cultural discourse. We shall be expanding upon these arguments below. In the meantime, the point is this: no matter how radical the contemporary postcolonial analysis, it can always be contained by virtue of its concurrence with established models which, for all their posturing and attitudinising, have been, and by and large continue to be, fundamentally apolitical and ahistorical.[3]

And yet, as even the most reactionary research cannot but attest, the apparently natural relationship between culture and politics has a history; it emerged and continued to exist only as the result of institutional and intellectual struggles over different ways of configuring the relationship between an imagined political community and the cultural forms through which that community could know itself. My contention in this book is that the specific location of those struggles was a form of *metacultural* discourse which has existed throughout modern Irish society in a wide variety of forms and locations, but which for convenience I shall refer to here as 'criticism'. That is, at the same time as Irish writers were 'performing' the nation in their creative texts – confirming or challenging the notion of a 'natural' Irish propensity for politicising cultural activity – Irish critics were engaging with the intellectual and institutional terms upon which such creative acts could sensibly and legitimately be made. In other words, a literature of resistance existed in dialectical relation with a diverse range of critical practices, the latter concerned with the establishment of a series of social spaces wherein such a literature and its particular effects could function and have meaning. Moreover, *after* culture becomes political and the Irish writers of the nineteenth and twentieth centuries grow self-conscious of their role as 'narrators' of the nation, criticism continued to play a crucial role in validating, refining and performing these narratives, and in facilitating or resisting the articulation of different modes of decolonisation.

Thus, the copula ('Culture *and* Politics') around which so much contemporary Irish and postcolonial debate is organised has a history of its own, a history characterised by the struggle between various ways of comprehending the relationship between cultural texts and the sorts of political communities from which such texts emerge. The main location of this historical struggle has been critical discourse. Therefore, the range of effects and practices comprehended by the concept of 'Irish culture' is delimited by a *metadiscourse* which purports to comment on, but actually constructs, the terms in which the 'original' or 'primary' discourse signifies. It becomes incumbent upon the modern critical community, I suggest, to acknowledge this situation and to set about the task of historicising the possibility of their own discourse. It is just such a task that I am advocating and attempting to practice here.[4]

This (re)turn to the question of 'the function of criticism' finds a resonance throughout much current intellectual debate, and two broad approaches have tended to dominate: one, textualist, particularly associated with deconstruction's establishment of writing as supreme metaphor for a post-humanist principle of undecidability and the revision of established philosophical hierarchies (such as literature/criticism); another, socio-historical, primarily associated with neo-Marxist and cultural materialist analyses and concerned to understand the role of cultural discourses in reproducing or contesting the organisation of 'society'. These aspects are inflected in particular ways in postcolonialist theory, and by way of expanding upon these introductory remarks I want to spend the rest of this chapter exploring these positions in a little more depth before stating once again my conviction regarding the importance of a critical history for a radical Irish cultural politics.

Criticism and Crisis

I noted in the previous chapter Homi Bhabha's indictment of the traditional forms of 'image analysis' which, he claims, have structured both colonialist and decolonising discourses. We saw also how this might feed into a general modular history of decolonisation. Bhabha's point, here and throughout his research, is significant in that it signals a shift in emphasis away from historical and cultural practices traditionally considered 'primary' – art, literature, or simply 'culture' – towards the systematic analysis of 'secondary' or 'critical' forms of discourse. This typically deconstructive manoeuvre is also characteristic of Spivak when we find her refusing to collude with the received opposition between 'original' and 'secondary' modes of discourse, or to privilege either the cultural text or the critical commentary – her own or that of any of the great modern 'founders of discourse' such as Marx, Freud or Derrida. For Bhabha and Spivak, as Robert Young has pointed out, 'the possibility of criticism comes not from experience or consciousness as in Said, but rather from the exploitation of a certain methodological scission within the rationalist project' (1990: 173–4). In intellectual terms, this is the moment of 'theory', signalling a willingness to confront the ideology which produces a range of social and institutional hierarchies and to consider (in this case) the ways in which such an ideology might be related to the histories of colonialism and decolonisation.

The move instanced by Bhabha and Spivak is entirely understandable, for anyone approaching decolonisation along the 'theory' route must be struck at some point by similarities between the kinds of oppositions and hierarchies discernible throughout colonial history (coloniser/colonised, active/passive, culture/nature) and the implicit hierarchy of 'primary' over 'secondary' discourses (culture/critique, literature/criticism, art/commentary) which structures modern intellectual discourses. Colonialism and criticism are in fact frequently invoked with reference to the same matrix of cultural structures, psychological assumptions and, in many cases, the same metaphors and linguistic tropes. In both instances, one element of an apparently 'natural' opposition is privileged, and in both cases this economy works by recourse to the same principles of originality, essence and presence.

The reasons for the links between colonialism and criticism lie in their shared roots in modern European thought. The concept of criticism began to perform important functions in the production and reproduction of certain characteristic modern modes of thought in Europe from the late seventeenth century. But it was during the eighteenth century that the principle of *critique* became an area of central concern for European cultural and philosophical thought, attracting interventions from many leading thinkers and writers. Thomas Docherty points to the origins of modern criticism in the Enlightenment and its objective to emancipate humankind 'from myth and superstition through the progressive operations of a critical reason' (1996: 480). Criticism was one of the principal mechanisms through which the rational, autonomous subject of Enlightenment reproduced himself as 'an agent of history rather than its victim' (481). Evolving from its classical roots in law, medicine and philology, critique emerged as the ultimate articulation of reason in an Age of Reason, the mode of discourse by which the great questions regarding humankind's existence could be posed and, hopefully, answered.

As with nationalism, however, criticism betrayed its Enlightenment roots in its radically ambivalent character. For at the same time as it was emerging as an important element of rationalist discourse, criticism's own status was coming into question, and thus was laid the basis for the constant crisis which has attended the discourse ever since. This crisis manifested itself as a stand-off between what Robert Con Davis and Ronald Schleifer describe as 'two different modes of critical analysis' (1991: 22). The first, exemplified by Kant, is a form of 'institutional' critique which attempts to discover 'the invariant conditions that govern the existence of any phenomenon. It subjects the actual to relentless questioning in order to discover sufficient reason why it should be so and not otherwise' (23). The second, exemplified by Hegel, is a form of 'transformative' critique which

> aims at criticizing 'positive' existing phenomena but not, as Kantian idealist critique does, simply to understand and make explicit the conditions and grounds that govern those phenomena and understanding itself. Instead, it aims to make something happen: to assert that things could be otherwise, that what exists is not necessary ... and that critique can allow us to imagine and to articulate a *narrative* in which things would be different from what they are ... if Kantian critique leads, over the course of the nineteenth century, to the study of the anonymous logical and linguistics systems that condition and govern abstract 'understanding', then Hegelian critique leads to the study of *particular* situations of understanding – discursive, psychological and philosophical – that condition and govern understanding. (25, original emphases)

Thus, modern criticism is from its inception working to different agendas, attempting to produce the Enlightenment subject as an agent of either Kantian or Hegelian narratives. As an example of this crisis (and one which bears upon future analysis), we might note the problems which beset modern English criticism from its inception in the early part of the eighteenth century and its subsequent implication in a fully fledged Enlightenment crisis. Literary criticism became important in England in the early eighteenth century as part of a quasi-emancipatory bourgeois discourse which attempted to locate power, knowledge and identity not in the absolutisms which had been responsible for the destructive enthusiasm of the seventeenth century, but in a new subject which found its authority in reason and its narratives in truth (Eagleton, 1984: 20ff.). English literary criticism confronted a problem, however, when it tried to reconcile this new historical subject with the traditional constitution of criticism as a *discipline* – a discipline both in the sense of a specific practice with its own internal laws, and in the sense of the right to police these laws and penalise their infringement. This sense of criticism as a discipline was received from the example of the ancients and medievals who had been primarily concerned with definition (Aristotle, Aquinas) and value (Plato and the Renaissance Neoplatonists). Thus conceived, criticism was a prescriptive and censorious discourse arrogating to itself the right to define a certain kind of social practice (literature) and to pass judgement on the validity or worth of individual instances of this practice, such judgements being frequently couched in a language which spoke metaphorically of the life or death of the author (Baldick, 1983: 7-9). In fact, 'criticism' still carries this implicit aggressive and pejorative sense as a remnant of its old disciplinary function.[5]

In Enlightenment discourse, however, definition and value are derived directly from reason, a non-hierarchical principle which claims to locate value democratically and evenly across universal humanity. Understood in this way, rationalism is a liberatory discourse, and it is indeed from its shared lineage with emancipatory reason that cultural criticism has long claimed a traditional radical and iconoclastic role. But such a notion left criticism having to reconcile the competing needs of the age of emancipatory rationalism with a discourse 'naturally' tending towards prescription and correction, what Eagleton calls 'the ineluctably negative moment of criticism' (1984: 20). And like its European counterparts, modern English criticism proved itself unable to resolve this central contradiction. For a short period in the early eighteenth century, reviewers like Addison and Steele attempted to be simultaneously humanistic and critical, to invoke emancipatory and censorious registers alongside one another. As Eagleton notes, however, critical discourse soon reverted to the kind of privileged and absolutist pronouncements which had characterised previous practices, the only difference being that the audience for such pronouncements was larger than had ever been possible before and growing all the time. Rather than demonstrating a rationalist principle of universal enlightenment, literary criticism in eighteenth-century England quickly became a policing exercise designed to demarcate an area of privileged activity, and to help constitute the subjects engaging in, and excluded from, this activity. The spaces and practices of literary criticism were appropriated by interested social groups seeking to recruit the discourse's ostensible communicative rationality for instrumental purposes. Thus, magazines and newspapers became more and more partisan until they were openly polemical; the essay and the review became tools of attack, and the promise of rationality and enlightenment initially held out by literary criticism was denied.

In fact, as Eagleton goes on to show, English literary criticism was fundamentally at odds with the age of reason and its peculiar social, political and historical narrative. Since the early eighteenth century, English criticism's status shifted from the security of having a message, a medium and an audience to the insecurity of social and intellectual marginalisation; from imagining a universal role in human affairs to becoming a specialist in one, not very significant, area of social experience. 'The contradiction on which criticism finally runs aground – one between an inchoate amateurism and a social marginal professionalism – was inscribed within from the outset' (1984: 69). Criticism was announced in the name of a liberatory reason and claimed a salient role in both the democratisation of culture and the emancipation of discourse. Yet, in its forms and strategies, it went on consistently to deny its own conditions of existence, emerging during the nineteenth century as a discourse, like nationalism, in a state of constant crisis. And like nationalism, criticism's crisis persists to the present. In this respect, F.R. Leavis emerges as the most representative modern English literary critic, an increasingly marginal figure making untenable claims for the role of literature and criticism in society while his lost audience carried on reading the 'wrong' material in the 'wrong' way for the 'wrong' reasons.[6]

It is possible to trace the evolution and trajectory of this crisis in particular intellectual and institutional forms, and in particular national formations, throughout the nineteenth and twentieth centuries. The history of criticism, that is, is informed by precisely the kind of oppositions and hierarchies to which deconstruction addresses itself. 'Deconstruction' as described by Con Davis and Schleifer, 'involves a reversal and reinscription of the usual patterns of interpretation' (1991: 166); that is, it works to expose what Derrida calls 'the indefinite process of supplementarity' (1976: 163) which is the characteristic mode of discourse in the West, and which, as interpreted by Christopher Norris, 'is precisely this strange reversal of values whereby an apparently derivative or secondary term takes on the crucial role in determining an entire structure of assumptions' (1987: 67). As one of the 'structure of assumptions' with which cultural commentators regularly have to contend is that which organises the relations between 'primary' cultural activities and their own 'secondary' critical practices, the logic of deconstruction, as Seán Burke points out, 'has led many poststructuralists to suggest that criticism itself has become a primary discourse' (1992: 159).

The received model of literary/critical activity which comprehends the latter discourse as 'secondary, parasitic, sponsorial' (161) cannot withstand a methodology geared towards the inversion and subsequent collapse of such well-established intellectual structures. Imre Salusinszky puts his finger on the matter:

> All of [the] favorite oppositions of criticism are open to deconstruction, but none more so than the really big, heavily invested ones that criticism has sought to establish between literature as a *whole* and all the things that it *isn't* ... not forgetting the daddy of them all, the distinction made *inside* criticism *between* criticism and its own object. I mean the inaugurating distinction between literature itself, made at criticism's degree-zero, only to be muddied at critical zero-plus-one by all those speculations about the 'literary' element in criticism, and the 'critical' element in literature'. (1987: 12, original emphases)

Literary criticism offers a particular case in point of the logic of supplementarity because it shares its mode of expression – textuality – with its object. Not only does literary criticism 'create' its own object (literature) by policing the terms in which judgements are passed over what is and is not to be included as 'literary'; at the same time, literary criticism, as Gérard Genette writes, ' ... is a metalanguage, "discourse upon a discourse"'. It can therefore be a metaliterature, that is to say, 'a literature of which literature itself is the imposed object' (1988: 63). When literary criticism attempts to represent something (the meaning of the primary text) it finds itself simultaneously creating something extra, something different (the critical text). The (secondary) critical text offers to *discover* the meaning of the primary text, but in so doing it produces another meaning which is specific to itself, in fact *covering* that text with its own evaluative and prescriptive languages (Barthes, 1972: 650).

The fact that it shares with its master text the same mode of signification, thus possessing potentially the same signifying capacity as a discourse it designates primary and different, is a profound source of embarrassment for literary criticism. The question deconstruction enables us to ask is: if a certain kind of writing is designated primary, replete, sufficient, intrinsically valuable, then why criticise it? The guilty secret always carried by literary criticism is that literature is not so primary or different after all; and that while the critical text must constantly demonstrate its own unworthiness in the face of its object, it simultaneously deconstructs the terms of that opposition by virtue of the very materiality of its discourse. Far from revealing the superior value of the artistic text, criticism merely highlights and stages that text's radical relativity, its inadequacy as a vessel for constant, universal meaning. Criticism attests to the fact that art needs commentary, the primary text depends on the secondary text just as much as literary criticism has traditionally been understood to depend on literature.[7]

At this point, literature's primary status *vis-à-vis* the critical text is in danger of fracturing under the weight of its own inherent contradictions. The whole structure of value and function depends on the simple narrative: first art, then criticism; but this narrative is constantly on the verge of collapse by virtue of the fact that criticism shares its mode of signification with its ostensible object. Critical discourse thus always operates to an alternative agenda, its major preoccupation being not with the *literary* text under scrutiny, but with the *critical* text under construction, not with the author and his or her *representation* of a certain social, political, and historical milieu but with the critic's construction of the same. If the gap between these two agendas becomes too wide (as it always does), breakdown ensues. Etymologically and discursively, therefore, 'criticism' implies crisis. The two words are in fact closely linked, so that whereas 'critical' carries the evaluative and analytic connotations employed here, it also registers, for example, in medical discourse, as something in need of immediate attention, as a crisis. It is in this context that Paul de Man states (in an influential essay entitled 'Criticism and Crisis'): 'the notion of crisis and that of criticism are very closely linked, so much so that one could state that all true criticism occurs in the mode of crisis' (1971: 8).[8]

It is a form of this deconstructive logic which Thomas Docherty activates when he notes 'a specific relation between criticism in modernity and the formulation of the emergent nation-state', his argument being that 'criticism as we know it depends upon an attitude which is, tacitly, nationalist in fact and origin' (1996: 479). Founded upon 'an anxiety about exteriority' (483), modern English criticism is characterised by 'the occlusion of the object of criticism in the interests of the production of what is fundamentally (if silently) a nationalist identity for the subject, the critic herself or himself' (479). As a practice, criticism is thus 'tied firmly to the place-logic of the nation-state' (493). In this case, the nation-state is England, and the critic's attempt to 'master' the cultural text engages with, and resonates in, political discourse as the (English) nation's attempt to master its (colonial, specifically African) Others. However, in the classic deconstructionist move, just as the critic works to contain the alterity residing within the text, he invariably discovers 'a covert form of the subject himself (Europe and not Africa)' (499). Thus, the trope governing the emergence and enduring character of modern English criticism since Dryden is tragic terror, 'a recognition of the self in the Other', and hence Docherty's somewhat fanciful conclusion 'that criticism has not even yet begun to happen for the simple reason that the object of criticism is constantly being circumvented in the production of a subject whose truth is guaranteed by autobiographical self-coherence (or subject-legitimation) rather than historical engagement' (500).

Docherty's dismissal of three centuries of literary criticism as essentially uncritical is, as we shall shortly see, untenable when it comes to an analysis of how a set of practices called 'criticism' has actually functioned in a range of societies over that same period. However, given the (albeit problematic) derivativeness of the liberal and radical models described in Chapter 1, it is a small step from the criticism/nationalism connection he proposes to the criticism/decolonisation connection which is the object of this study. Even at this early stage, one has only to think of some of the influential figures from modern Irish cultural history – Edmund Burke, Samuel Ferguson, Thomas Davis, Standish O'Grady, Oscar Wilde, Douglas Hyde, W.B. Yeats, Daniel Corkery, Seán O'Faoláin – to appreciate not only the intellectual importance of critical discourse but its implications for the other well-known issues pre-occupying these figures.

The Institution of Criticism

If criticism needs to be located within the narrative of theory – that moment when 'secondary' discourses of commentary and critique become self-aware and begin to reflect upon the hierarchies within which they have traditionally operated – there is another narrative, institutional and historical, within which it also needs to be placed. For if modern critical discourse serves to reveal textuality's inherent phenomenological crisis, it seems clear that it is also a highly specialised function occupying specific social and institutional spaces, enabling special individuals to perform special kinds of social and political tasks. The category of 'criticism' is in fact a problematic designation for a highly diverse set of practices occupying certain social and political spaces wherein a discourse of commentary operates. This is not to deny the force of the deconstructive moment examined in the previous section. But that crisis needs to be contextualised so that its specific local and material effects may be calculated. As Peter Uwe Hohendahl argues: 'the legitimation crises of criticism are long-term problems, ultimately rooted in the structure of the sociocultural system and its relation to the economic and social systems' (1982: 42). In the present instance, the particular sociocultural system within which the crises of criticism are to be analysed is that of Ireland during the period of its decolonisation. The critical act, like the cultural act which it ostensibly serves, is a social act; criticism is a body of rhetorical strategies waiting to be seized and used. One of the aims of this book is to look at the ways in which criticism has been seized and used as part of the process of decolonisation in modern Ireland.

The history of literary criticism constitutes a respectable subcategory within the institution of literature, and, as far as the anglophone tradition is concerned, there has been no shortage of such histories ever since George Saintsbury's *A History of English Criticism*, published in 1911 (extracted from his earlier *A History of Criticism and Literary Taste in Europe* published in three volumes between 1900 and 1904). It is no surprise that Saintsbury's book appeared at a time when English as a university subject was consolidating, for such a discourse may be understood as part of a process of disciplinary self-constitution, an attempt to demarcate the boundaries of literary activity and, with its characteristic modes of taxonomy and categorisation, to dispel the aura of dilettantism which trailed from the nineteenth century.[9] An established, evolutionary *narrative* of literary criticism would add to the notion of English Literature as a discrete, valid discipline. Thus, a distinctive *metacritical* discourse has emerged during the twentieth century as an important element of the literary edifice alongside history, criticism and aesthetics.

Much of this traditional metacriticism, however, is characterised by what one observer has called a 'dignified vacuousness' (Pinkney, 1982: 248), concerned not so much with an analysis of the historical factors bearing on the emergence of criticism as with the protection of a highly specialised definition of the 'literary'. In one relatively recent example, George Watson's *The Literary Critics*, despite the author's attempt to break out of what he calls 'the Tidy School of critical history' which assumes 'that what we call literary criticism is, with some embarrassing examples, a single activity, and that its history is the story of successive critics offering different answers to the same questions' (1964: 10), what he actually offers is a variation on this method. *The Literary Critics* is essentially a catalogue of interventions by 'great men': 'The great critics', he claims, 'do not contribute: they interrupt' (11). Watson describes a pseudo-taxonomy of certain kinds of 'criticism' (legislative, theoretical and descriptive, focusing predominantly on the last) but ignores the social and institutional constitution of the discourse as well as the many other spaces and practices throughout modern British history in which principles of cultural commentary have been active. By focusing upon a particular kind of textual commentary, Watson attempts to prop up the established model which sees cultural discourse organised along specific hierarchical lines. In doing so, however, he ignores the dialectical relationship between the critical and the literary, and the ways in which the latter is a category contingent upon the material fortunes of the former.

Against this essentially unhistorical approach, modern Marxist theory has been concerned, in the words of Terry Eagleton, 'to pose the question of under what conditions, and for what ends, a literary criticism comes about'. He goes on:

> [Criticism] does not arise as a spontaneous riposte to the existential fact of the text, organically coupled with the object it illuminates. It has its own relatively autonomous life, its own laws and structures: it forms an internally complex system articulated with the literary system rather than merely reflexive of it. It emerges into existence, and passes out of it again, on the basis of certain determinate conditions ... It is the history of *criticisms* which is at issue. We are seeking the determinants of the particular historical 'spaces' which make the emergence of such an object possible in the first place. The science of the history of criticisms is the science of the historical forms which produce those criticisms – criticisms which in turn produce the literary text as their object, as the 'text-for-criticism'. (1978: 17).

The promise of a 'science of the history of criticisms' belongs very much to the Althusserian moment of European Marxist theory which began to make itself felt in England during the 1970s. In a later, more humanist mode, Eagleton used Habermas's notion of 'the public sphere' to assess literary criticism's part in 'the social production of forms of subjectivity' (1984: 124) in modern English history. From a similar perspective, Peter Uwe Hohendahl complained that in the case of what was then West Germany, 'The common knowledge that a literary text is embedded in a historical context that can be defined in cultural, political and social terms has not been fully appreciated in the examination of various forms of literary criticism – scholarly books and articles, journalistic essays, book reviews in newspapers, and the like' (1982: 11). He then lists the sorts of issues which should concern a properly materialist study of the institution of criticism, including 'the social role of the critic ... the social organization of criticism (the press, associations, academies) ... the connections of these organizations to the institutions of the whole society (state, church, court, parties)

... and the significance and function of literary criticism in the system of the whole society' (236). In both accounts, cultural and social realms are fused through the mediating category of the public sphere, 'a construct whose function is to make the dynamic processes between the spheres of society, state, and culture describable' (235). In this way Eagleton and Hohendahl attempt to side-step the 'reflection' models of vulgar Marxism and create the conditions for a highly sensitive analysis of 'the social models which guide and control the activity called literary criticism' (Hohendahl, 1982: 12).[10]

Even such relatively sophisticated models are not enough for some theorists, however. In what must count as the most rigorous materialist analysis to date, Tony Bennett dismisses most of the modern assessments of the function of criticism (including many influential Marxist accounts) as still mortgaged to a totalising bourgeois-humanist model. Such a model, he claims, fosters a notion of criticism as a 'practice which constitutes literary texts as the sites for a totalising commentary which aims at the ethical-cum-political transformation of subjects without enquiring into the historically specific conditions which have put such texts into the place – discursive and institutional – where they can be so constituted ...' (1990: 205). Bennett's point is that even those concerned to expose the networks of power within which literary criticism has traditionally operated tend to accept the practice as a specific kind of operation which the critic performs upon the text with a view to altering the perspective of the reader in particular ways. But this, he claims, represents just one, privileged way in which the relations between texts, readers and practices of textual commentary have been historically organised. It is incumbent upon any radical history, he suggests, to subject the practice itself as well as its effects to materialist analysis if it is to avoid complying with fundamentally anti-materialist systems of thought. Only by examining 'the more specific, multiple and differentiated political questions posed by the differentially constituted institutional and discursive domains in which literary texts are deployed and practices of textual commentary effectively engaged' (219–20) may the critical historian appreciate 'what roles might be performed by different types of critical practice given the varied institutional domains, and their varied publics, in which such practices are operative' (242). For Bennett, then, the history of criticism is primarily a question of social and institutional deployment rather than any radical interpretative method – new or rediscovered – which might be brought to bear upon the text.[11]

Such an ultra-materialist approach also accounts for Bennett's suspicion of a 'counterpublic sphere' upon which many oppositional theorists and commentators base so much of their hopes for a renewed radical criticism.[12] The nostalgia which haunts many of the so-called 'radical' accounts of the function of criticism for a permanent oppositional site housing a stable critical politics locked in battle against the bourgeois state and its regimes of truth, he suggests, flies in the face of a truly radical and effective cultural politics. The idea of a singular function for a singular notion of criticism is in itself flawed; however, 'If a brief for criticism is called for ... it is one that will enable critical practices to operate multiply and variably on the sites of such contradictions (of the state) rather than – in constructing a totalising contradiction of its own – outside and independently of them' (236).[13]

Although broadly sympathetic to the various historical materialist approaches outlined above, I perceive no natural repugnance between them and the deconstructive method broached in the previous section. As I understand it, criticism articulates a crisis that has significant ramifications for the European Enlightenment subject, but that subject ranges over a complex network of interlocking spaces and temporalities which determine the specific effects such a crisis will have. This is the theoretical model, sufficiently broad in scope and eclectic in method, I wish to deploy throughout the remainder of this study. At this point, however, it is necessary to pose the question of the specific relationship between criticism (understood now as an array of dispersed practices and effects rather than a singular function or process) and decolonisation; that is, to consider the ways in which a constellation of critical practices and functions has worked to constitute and contest the formation of a particular historical model of the nation-state.

Criticism and Decolonisation

Spectre-like, 'the function of criticism' has returned to haunt modern intellectual debate. Systematically repressed in the Western cultural consciousness, there are in fact pressing cultural and political reasons for this return. Criticism, critique, commentary: these and related terms need to be salvaged from the hierarchical structures in which they have been traditionally positioned and seen instead as elements of the discourse in which both colonial and decolonising strategies gain their force and their coherence. Ever since Plato decided to expel the imaginative writer from his ideal state, criticism's self-professed secondary and revelatory role has masked a much more active and engaged agenda. The philosopher's concern in *The Republic* was with the health and survival of the state; to those ends, 'bad' art was to be censored, 'good' art to be tolerated, and a critical discourse invented to enable those in power to tell one from the other. Thus, at the dawn of Western philosophy we find criticism performing unashamed ideological tasks as well as being appointed the moral guardian and social conscience of a certain kind of imagined community.

Criticism's inherent political dimension has been maintained and enhanced during the modern era. From the perspective of the late 1990s it has become possible to configure the relationship between criticism and decolonisation as a form of allegory or metaphor of the relationship which, in most of the received accounts, obtains between culture and nationalism – that is, as the same basic formal and conceptual preoccupation of creative artists with the imagined community, carried on at another remove. On the other hand, one can follow the revisionist path and seek to dismantle what is seen as a disabling and confining relationship between culture and decolonisation – that is, refuse the criticism/politics nexus as merely a metadiscursive rehashing of the same old tired nationalist narrative to the exclusion of alternative models such as socialism, feminism or cosmopolitanism. Both these models seem flawed to me in that each maintains an implicit dualism in which criticism and decolonisation are considered as categorically separate discourses. I want to suggest a much more intimate and enabling relationship between criticism and decolonisation, arising from their problematic engagement with the characteristic discursive modes of the Enlightenment and manifested throughout modern Irish history in a range of social practices and institutional effects.

Hohendahl argues that 'the paramount task of the nineteenth-century critic' was to define 'the national cultural identity', and this is turn 'was closely related to the problem of political identity' (1982: 15). But in Ireland the relationship between criticism and forms of colonising and decolonising politics can actually be dated at least a century earlier. When Irish subjects began to formulate cultural responses to colonialist practices in the years after the Treaty of Limerick, forms of historical and literary criticism became crucial in the production and contestation of colonialist representations (Leerssen, 1996a: *passim*). The critical text, rather than merely serving a pre-existing cultural text, became a primary location for the hegemonic encounters between colonialism and the various modes of decolonisation examined in Chapter 1. This situation is obviously complicated by the presence of a settler community attempting to negotiate a dual role *vis-à-vis* the remnants of Gaelic Ireland and an increasingly unsympathetic Mother country. As we shall see in the next chapter, however, the critical controversies surrounding the status of texts discovered, created and translated during the First Celtic Revival from the mid-eighteenth century on is one clear location for the formulation of discourses of resistance. The business of such a critical discourse was to create a series of social and institutional spaces in which a range of cultural (and thus political) effects could function and have meaning.

Hohendahl goes on to state: 'Traditional literary histories and surveys have paid notably little attention to the participation and significance of literary criticism in literary life. To put it positively: a future history of literary criticism should be conceptually integrated into the history of literature, and its functional value should be established' (1982: 226). My contention here is that this criticism/decolonisation connection constitutes a fundamental aspect of the modern Irish 'cultural' imagination, and that, at least since the late eighteenth century, the debate surrounding 'the function of criticism' has also always been a debate about the function of the nation and the relations between colonising and decolonising subjects. Thus, when we turn to Irish culture of the past two centuries to observe the emergence of forms of resistance to colonial domination, it is necessary to establish the 'functional value' of critical discourse and to 'conceptually integrate' it with the imaginative, artistic and creative discourses which have dominated the analysis of modern Irish cultural history. The acts of literature in which the Irish nation is 'performed' throughout the nineteenth and twentieth centuries have themselves to be performed in prefiguring and/or retrospective critical discourses, discourses which in their own characteristic forms and practices engage with the various modes of Irish decolonisation. Despite the claims of generations of both imperialist and nationalist critics, that is, Irish culture cannot express, reflect, embody – or any of the other favoured metaphors – the decolonising nation until it is so constituted by an enabling metadiscourse: criticism.

So far I have been discussing the development of criticism in more or less abstract terms and in a predominantly English framework. The question of how Irish decolonising critics negotiated a metropolitan element apparently structured into the discourse and tried to re-articulate it for their own redemptive ends will be the subject of the next chapter, while the various strategies instigated during the 1950s to renegotiate the terms of the criticism/nationalism debate will be examined at length in Part 2. If, as I am arguing, 'the Irish literary tradition' is a strategic category constructed in the Irish critical imagination, then there would appear to be serious implications both for contemporary artistic practice and for contemporary cultural criticism.

3 Critical Encounters

Having delineated the theoretical and methodological bases for an analysis of modern Irish literary criticism, in this chapter I wish to examine some representative examples from the past two hundred years by way of an introduction to the main study offered in Part 2. The method here will be to compare and contrast a number of more or less contemporaneous critical interventions from four loosely defined moments of modern Irish history, such 'encounters' embodying certain discursive characteristics which enable us to consider them as examples of the liberal and radical modes of decolonisation described in Chapter 1. Liberal decolonisation, to recall, refers to a form of resistance in which the subordinate subject – native or settler – seeks *equality* with the dominant colonialist subject and is concerned to 'raise' the status of the colony/nation up to that of the imperial centre. Radical decolonisation entails the use by the subordinate subject – native or settler – of his differential status as the basis for both a repudiation of metropolitan values and a concomitant celebration of a national identity characterised as unique and authentic. These modes have been defined as *limited* but *necessary* responses to the fact of colonial domination. At the same time, we shall be looking to identify modes of critique which 'problematise, even as they engage with, the dominant narratives of decolonisation'.

Neo-classicism and Celticism

Ireland in the eighteenth century witnessed a classic colonial encounter between three communities – native Gaelic, settler Anglo-Irish, offshore English – engaged in hegemonic activity as part of the battle for political and cultural leadership. Despite the rout of 1688-91, the ghost of an old Gaelic civilisation survived behind a wall of legislative and cultural apartheid. As the century progressed, the Irish-speaking Catholic population recovered slightly but not so much that it ever exercised more than a marginal influence on political decision-making or cultural trends. The Penal Code which had been introduced in the years after the

Treaty of Limerick in 1691 rendered the Catholic Gaelic-speaking majority non-citizens, and the legal and cultural battle for 'Irish' freedom was, until the end of the eighteenth century at least, undertaken predominantly by the relatively small Anglo-Irish population. This community learned to consider itself an independent Protestant nation fully capable and deserving of autonomy from England. Anglo-Irish 'Patriots' argued that the Old English colony which they had inherited in fact constituted the Irish nation and that this longevity entitled them to all the rights and benefits claimed by freeborn Englishmen under Magna Carta. Taken together, the thinking of Swift, Burke and Grattan contributed to a notion of Ireland as an associated yet distinct kingdom which, given the respect it deserved, would remain loyal to a non-coercive English sovereign.[1]

The 'Protestant Nation' appeared to have arrived in May 1782 (ratified by British parliament on 22 January 1783) with the recognition of parliamentary independence, famously signalled by Grattan: 'Ireland is now a nation. In that character I hail her, and bowing in her august presence I say *Esto perpetua*' (quoted in Curtis, 1961: 315). But which Ireland? 'Anglo-Ireland' had always been a problematic category in that it was composed of two apparently antagonistic identities. The *Anglo*-Irish subject had access to the practices which would assure dominance over Gaelic Ireland; by the same token, the Anglo-*Irish* subject was hegemonically subordinate *vis-à-vis* his English neighbour. This situation fuelled 'a curiously schizoid identification' (Foster, 1989: 178), and helps to account for the insecurity of the Anglo-Irish during the period of their so-called 'Ascendancy'. It was always possible that the arguments with which 'Patriots' opposed England could be appropriated and used against them, providing Gaelic Catholic Ireland with an example which it might follow at the expense of the Anglo-Irish themselves (Leerssen, 1996b: 8–32). Paradoxically, the 'victory' of 1782 precipitated a crisis for the settler community, a crisis which may be discerned in Anglo-Irish critical discourse in the period after the formal constitution of 'Grattan's parliament'.

The discourses with which the Anglo-Irish critic might engage during this period partook of the intense debates which had riven European culture throughout the century. These in turn fed into different models of the national community. We recall from Chapter 1 that Anthony D. Smith described 'two very different kinds of national political identity and community' (1991: 99): the first, a territorial model based on the principles of rationalism, equivalence and classical order; the second, an ethnic nationalism founded on indigenous romanticism, medievalism and the idea of difference. Smith dates 'a new quasi-Grecian taste' (87) and the influence of Republican Rome in European society from the 1760s. These trends were an extension of the 'Quarrel of the Ancients and the Moderns' which had animated cultural discourse in the advanced European formations around the turn of the eighteenth century. At the same time, running parallel to classical nationalism was another impulse, irredentist and redemptive, geared towards the confirmation of an authentic, collective, ethnic identity. Refined by German Romanticism, 'the ideal of autonomy gave rise to a philosophy of national self-determination and collective struggle to realize the authentic national will – in a state of one's own. Only then would the community be able to follow its own "inner rhythms", heed its own inward voice and return to its pure and uncontaminated pristine state' (76–7). Smith continues:

> (although) variants of a wider romanticism, a yearning for an idealized golden age and a heroic past that can serve as exemplars for collective regeneration in the present ... Yet the opposition between enlightenment and medieval romanticism also mirrors a deeper cultural and social cleavage between the two ethnic bases and routes in the formation of nations, from which two radically different concepts of the nation have emerged. (91)

These trends found their way into English critical discourse in the second half of the eighteenth century as a debate between neo-classicism and Celticism (Atkins, 1951; Snyder, 1923; Wimsatt Jr and Brooks, 1957). Neo-classicism had moved away from the strict regimentation introduced by French critics such as Boileau, Rapin and Bossu, but was still concerned with classical precedent (even if the major influence was now Longinus rather than Plato, Aristotle, Virgil or Juvenal). Thomas Docherty, we recall from Chapter 2, insists that English formal criticism was from its inception characterised by 'the occlusion of the object of criticism in the interests of the production of what is fundamentally (if silently) a nationalist identity for the subject, the critic herself or himself' (1996: 479); while Terry Eagleton (1984) has argued that critical discourse in England during the eighteenth century was crucial for the emergence of an enlightened bourgeois public sphere, itself part of the process whereby a modern English identity was consolidated. For the English critic, then, engagement with neo-classical critical discourse was both a means of stressing English participation in the essential continuity of European civilisation and also a way of marking the distance between developed communities (such as

England) and those benighted populations (as in Ireland) still labouring in the dark ages.

Celticism grew out of interest in Hebrew and European vernacular traditions before becoming a major medievalist cult, especially in England (Snyder, 1923). Rather than mitigating the hegemonic claims of neo-classicism, however, the emergence of interest in Celtic culture and history in the British Isles was part of a project of identity formation dictated from the metropolitan centre to the Celtic margins. At the same time this discourse attempted to institute a wider 'British' ideology dominated by the English centre but drawing consent from its constituent peoples. As Declan Kiberd has argued: 'The notion of "Ireland" is largely a fiction created by the rulers of England in response to specific needs at a precise moment in British history'(1985: 83); and elsewhere 'Irish intellectuals deduced that the intent of English policy was straightforward: to create a "Sasca nua darb anim Éire" (a new England called Ireland)' (1995: 15). In similar vein, Luke Gibbons writes that '"Celticism" ... was an attempt by a colonial power to hypostasize an alien, refractory culture in order to define it within its own controlling terms' (1991: 568).

As members of a community looking to constitute its own identity in relation to Gaelic Ireland and England, both these models were available to Anglo-Irish critics, demanding different methodologies and different texts-for-criticism. Criticism engaging with the neo-classical sphere of influence was concerned with the identification through close reading of the primary text's balance of decorum, style and emotion. At the same time, the critical text must also demonstrate the same qualities; 'the stylistic idea for prose and poetry alike' as one literary historian explains, 'was the Horatian combination, elegance united with functional efficiency'(Rogers, 1978: 12). In contradistinction, Celticism was best served by activities and techniques (such as philology, history and archaeology) which were realist, historicist and populist in impulse. Literary medievalism, as Smith describes it,

> depended, by its method of inquiry and its literary data, on the records of specific peoples and milieux in order that earlier periods of a community's history and culture were to be reconstructed as they had really existed ... medievalist literary historicism provided the concepts, symbols and language for the vernacular mobilization of demotic *ethnies* and a mirror in which members could grasp their own aspirations as they took shape amid the transformations wrought by the Western 'revolutions'. Here they could read of themselves as a unique community with a peculiar 'genius' and distinctive culture, and recognize a 'national

> character' that demanded its autonomy so that it could live authentically. (1991: 88, 89, 90)

Neo-classicism and Celticism, therefore, signified different concepts of the national community. In terms of the modes of decolonisation described in Chapter 1, critical engagement with the former constituted a 'liberal' resistance in that it attempted to 'raise' Irish cultural activity onto a level with English practice. Celticism, however, was a 'radical' decolonising activity in as much as it attempted to assert an authentic Irish identity separate from English influence. As we shall now go on to observe, however, no overall strategic advantage ensues from embracing English letters rather that Irish antiquities as the site of the critic's decolonising activity; both modes remain subject to the limitations attending any discourse of resistance cast in the oppositionalist terms received from the dominating formation.

William Preston's essay 'Thoughts on Lyric Poetry' was first delivered as a paper before the newly formed Royal Irish Academy (RIA) on 11 December 1786.[2] The RIA was a group of mainly Protestant gentlemen scholars which had grown out of earlier attempts (such as the Dublin Society, founded in 1731, and the Hibernian Antiquarian Society, founded in 1779) to institute a learned society in Ireland along the lines of the Royal Society in England. Engagement with such a discourse was a highly ambivalent activity for any Anglo-Irish subject, however, for there was a constant danger that the more they demonstrated an ability to lead Ireland in matters of culture and taste, the more they acknowledged their deference to a 'natural' English superiority. From the first the members of the RIA were aware of their strategic disadvantage in comparison with their English fellow academicians:

> To the several advantages which Europe has within these latter centuries experienced from the cultivation of science and polite literature, this kingdom unfortunately has remained in a great measure a stranger. As no Irishman's partiality will deny this, so no man's prejudice should be suffered to make it an occasion of illiberal imputation on the capacity of Irishmen, while in the state of the country so many local peculiarities may be found fully sufficient to account for it. (Anonymous, 1787: ix)

Already we find the academicians on the defensive; the exoticism of the 'stranger' and the 'local peculiarities' at once mark these Irishmen off as different from the rest of Europe and as the potential victims of a normal 'man's prejudice'. At the same time they do not refuse the local nomenclature of 'Irishmen', for to do so would delegitimise their claim to the national leadership. There is thus a tension between wishing to disown the badge of subordination (Irishman) and desiring to demonstrate their ability for cultural and political leadership by trafficking in valuable cultural currency to which they, as particular kinds of (Anglo-) Irishmen, have access.

The contributors to the RIA's first volume were dealing with some of the major questions of eighteenth-century English literature – sublimity, style and, in Preston's case, classical precedent.[3] Even as the choice of 'Polite Literature' immediately signals Preston's acknowledgement of English cultural leadership, we find him engaging with a strain of *anti-classicism* which had been growing in English critical discourse throughout the century (Atkins, 1951: 146–224). Anti-classicism was influenced by the concomitant rise of literary medievalism, and also by the notion of a 'natural' English moderation responding to the *ultra*-classicism introduced by the French critics and their misguided English apologists. Preston attacks an earlier essay in which the English poet and dramatist William Mason had occasion to find fault with some of the work of his friend and mentor, Thomas Gray.[4] The substance of this attack is that Gray was too much given to the irregular ode, a form which encouraged 'wild and jejune' sentiments and a wayward and potentially dangerous imagination. Instead, Mason maintains (in typical neo-classical fashion), the poet should adhere rigorously to the classical Pindaric rules of *strophe*, *antistrophe* and *epode*, which will allow the poetic sentiments to 'live' and will encourage regularity of thought and habit in the reader.

The Anglo-Irishman Preston, on the other hand, argues in favour of irregularity, accusing those who adhere to the classical code of 'pedantry' and 'servile imitation':

> The mere regular return of an uniform stanza, if that stanza does not afford a copious interchange of melodious sounds, is not a work of much difficulty in the execution, or merit in the perusal; neither can it be said to impose any very strong, at least it does not impose any very useful curb, on the wayward imagination; nor will it, I presume, be found a very effectual means of excluding compositions wild and jejune: In truth, I am inclined to doubt whether this desirable end can be obtained by the adoption of strophe, antistrophe, and epode. It would be invidious to quote particular instances, but any one who will take the trouble of turning over some of our miscellaneous collections, and other books of modern poetry, will find things called odes, which are at once wild and jejune, though trimmed and laced up in the straight waistcoat of strophe, antistrophe, and epode, according to all the severities of the Greek masters. (in Anonymous, 1787: 60)

Preston calls for a form of verse which would be free of the classical fetters which, he claimed, were threatening to smother the true spirit of English poetry. For proof of these claims he evokes certain properties – such as 'primitivism' and 'nature' – which constituted part of the discourse of literary medievalism and Celticism:

> I have not a doubt within my mind of the irregular ode being the first form of composition adopted by mankind, in their first wild attempts at literature. Poetry has ever been the delight of men in the first stages of society: the earliest recitals of events among them have been in verse ... The first literary production, in an unpolished nation, where the pure dictates of nature prevailed, was a poem, and that poem an irregular ode ... and I am confirmed in my opinion, by finding that several specimens of the antient poetry of uncivilized nations bear this form. (73)

Thus, by linking anti-classicism, primitivism and nature, Preston produces a version of poetic discourse which celebrates poetic freedom of form and expression. He is not arguing in favour of the 'wild and jejune' or the unfettered imagination; it is rather that he rejects the belief that proper poetic effects can be achieved through adherence to a false doctrine of classical regularity. These effects are in fact the same ones towards which the classical and neo-classical artists aspire; it is only the manner of achieving them that Preston is disputing:

> A correspondence of the sound with the sentiment is certainly a very great beauty, and the poet should endeavour to obtain it, whenever it may be had, without sacrificing more important things ... Now, I believe it cannot be denied ... that a free stanza, which may be varied at will, and made light and airy, slow and plaintive, or swelling and sonorous, according to the subject matter, will give the poet a much better chance of attaining this excellence, whatever may be its value. (71)

Decorum is not at issue, only the means of achieving it; one can apparently achieve regularity ('correspondence of the sound with the sentiment') through irregular poetic practices. In similar fashion, the actual form of Preston's essay seems to belie the argument against regularity which is its subject, for it is in fact a highly structured and articulate piece of prose, fully regular in a neo-classical sense.[5] The text, in fact, is typical of the transitional phase between two distinct cultural paradigms – neo-classical and romantic – and this 'transitional' status precisely rehearses the situation of the Anglo-Irish *vis-à-vis* their English and Gaelic Irish cultural/political competitors. The argument for irregularity leaves Preston tactically exposed to English hegemonic claims which are evoked in a discourse of Celticism specifically adapted to confirm English domination. By the same token, the *form* and the *context* of the argument – the language, the structure, the cultural codes and institutional powers with which it engages – signify Anglo-Irish acknowledgement of English cultural leadership. Yet these same elements – Celtic irregularity, English form – also signal Anglo-Irish cultural and political dominion over Gaelic Ireland. So the very discourse which renders Preston subordinate to one group renders him master of another.

In *Historical Memoirs of the Irish Bards*, first published in 1786, Joseph Cooper Walker opts for an 'Irish' subject matter, but goes on to engage with this subject matter through certain recognisable English cultural codes.[6] As an example of 'the coloniser who refuses' Walker's book is 'radical' in as much as it attempts to identify the 'inner rhythms' through which Anglo-Ireland might assert its difference from England.[7] The choice of subject matter, however, exposes Walker to the discourse of Celticism, that 'fiction', to reprise Kiberd, 'created by the rulers of England in response to specific needs at a precise moment in British history'. At the same time, the critic is enabled to demonstrate his superiority over Gaelic Ireland through the very materiality of his text, his ability to describe, contain, and thus have power over that section of the population. In both cases, the distribution of power remains the same, but with one important qualification: Walker's 'radical' discourse, consciously or unconsciously, produces something which escapes his ability to describe and control, a space for 'Irishness' which cannot be contained by his own hegemonic discourse. This space, although of very little contemporary consequence for the Gaelic Irish, having once been broached would be available at later stages as a powerful weapon of resistance to colonialist discourse, and in the hands of certain decolonising intellectuals of the nineteenth century would in fact become the basis of an alternative hegemonic discourse founded on the principle of Gaelic Irish exclusiveness and priority.

With *Irish Bards* Walker was intervening in a discourse which in the 1780s was nearing saturation point.[8] Throughout Europe, antiquarianism was an incredibly popular and highly charged cultural activity which reflected the contemporary fascination with the past, as evidenced by the work of intellectuals like Rousseau, Herder and Winckelmann. Irish antiquarianism was dominated (excluding the brilliant Charlotte Brooke) by a few gentlemen scholars – Protestant and Catholic – who formed the basis of the Antiquities section of the RIA: Sylvester O'Halloran, Charles

Vallancey, Charles O'Conor and Walker himself.[9] These Irishmen, however, were not writing of universal human history as was Rousseau, or a shared racial history as was Herder, or an admired foreign history as was Winckelmann, but a specific, ongoing colonial history in which they and their cultural interventions played a vital role. As Smith argues:

> The uses of ethno-history were essentially social and political. Nationalists were interested not in inquiring into 'their' past for its own sake but in the reappropriation of a mythology of the territorialized past of 'their people'. Throughout, the basic process was one of vernacular mobilization of a passive *ethnie*, and the politicization of its cultural heritage through the cultivation of its poetic spaces and the commemoration of its golden ages. (1991: 127)

In *Irish Bards* Walker is certainly concerned with the 'cultivation' and 'commemoration' of an 'Irish' golden age as part of the project of resisting English cultural hegemony. However, controlling a discourse based on an identity which was both dominant and subordinate by virtue of the same quality of 'Irishness' proved difficult for Walker. The more he claimed the Irish nomenclature, the more he was anticipated and undone by the discourse of Celticism in which he was looking to intervene. The difficulty he and his fellow Anglo-Irish antiquarians faced is neatly summed up in this statement from the Preface to Volume II of *Irish Bards*:

> It was hinted to me by a friend, who perused my manuscript, that I dwell with too much energy on the oppressions of the English; treading, sometimes, with a heavy step, on ashes not yet cold. But however thankful for the hint, I cannot subscribe to his opinion. I have only related unexaggerated historic truths. This was my duty, and from this duty no mortal frown can make me swerve ... But the wrongs of the English only live now in the page of history. Mingling their blood with ours, that brave people have conciliated our affections. We have taken them to our arms, and stifled the remembrance of their oppressions in a warm embrace. (1818: 3–4)

Two different identities operate on either side of the 'But': before, it is the Anglo-Irish intellectual attempting to interpellate and lead Gaelic Ireland by virtue of his right and ability to represent Ireland; after, it is the Anglo-Irish intellectual trying and failing to ward off the implications of Celticism. Walker's fate is that of the typical Anglo-Irish settler, inhabiting a contradictory discursive site that is ineluctably Irish while simultaneously greater and lesser than that designation.

Walker announces his superiority over Gaelic Ireland and his ability to contain and represent that category through the employment of the metropolitan 'plain' style encouraged by neo-classical literary theory, and also by utilising a plethora of source materials in an effort to give his work that semblance of scholarly authority which will gain him credibility in metropolitan intellectual circles.[10] The scepticism he shows towards Macpherson, for example, is simultaneously an exercise in Anglo-Irish self-fashioning and implicit acknowledgement of English cultural leadership; that is, from the strategic position of a dominant Anglo-Irish subject, Walker arrogates the right to say what is and is not a culturally (and politically) legitimate experience, whereas from the point of view of a subordinate Irish subject he fully consents to, and is implicit in, English cultural leadership:

> Though Cucullin flourished about two hundred years before the reign of Cormac, Mr. Macpherson has made him contemporary with Fin, whom he calls Fingal ... Mr. Macpherson always changes ALMHAIN into ALBAIN, that is Scotland: for M and B are commutable in the Gaelic or Iberno-Celtic language, a circumstance of which he takes advantage. (51–4)

The style of this passage (indeed of the whole volume) and the sorts of knowledge it professes enables the author to assume a dominant position with regard to his material, an authority which resonates in contemporary political discourse. In anecdote after anecdote, footnote after footnote, 'sources' are employed to validate and reproduce Anglo-Irish hegemony over a textually constituted Celtic identity. The style, however, merely confirms Walker's subservience to English cultural values, while the scholarship is, for the most part, spurious. One obvious blindspot is the author's insinuation of Anglo-Ireland into an impossible role in pre-invasion Irish history. More generally, there is a discrepancy between the authoritative tone of the text and a lack of reliable primary sources. Walker admits 'that my knowledge of the Irish language is so very confined' (xvi) and has to rely on genuine bilinguists like Charlotte Brooke and Theophilus O'Flannagan for the interpretations and translations on which he bases his claims. There is in fact very little original research in *Irish Bards* and most of Walker's claims are based on secondary sources, some of which even at the time were notorious for their unreliability and their absurd Celtomanic claims.[11] Thus the text is peppered with provisos and qualifications along the lines of 'Hence, we may conclude' (77); 'in all probability ... it only remains to be admitted' (102); 'we may venture to suppose ... It is no wild conjecture' (116); 'May we not suppose' (120); 'we may infer from thence' (125); 'We have good reason to believe' (135), and so on.

What the text lacks in quality, however, he tries to make up in quantity, with the main argument threatening to disappear beneath a mass of annotated material the relevance of which is mostly unclear, the authority of which is mostly questionable.[12] The style, also, often threatens to break down under the welter of clauses and subclauses, with the comma and the semi-colon employed copiously and gratuitously. The strategic decision to adopt a self-conscious metropolitan style in relation to his Celtic material implicitly acknowledges English cultural hegemony; the improper execution of that style merely disables Walker and leaves him exposed to the hierarchical discourses (neo-classicism and Celticism) which ordered the cultural relations between coloniser, settler and colonised.

The main problem was that antiquarianism was still in its infancy in the eighteenth century and its research techniques were not yet equal to the primary sources of Irish history. Until they became so, thanks to the work of nineteenth-century scholars, antiquarianism would remain the province of speculative amateurs like Walker himself. In the meantime, antiquarian discourse functioned to reproduce an Anglo-Irish identity which was simultaneously dominant and subordinate, for while the words on the page remained the same, the text offered various reading experiences to various strategically located subjects. It was the fate of the Anglo-Irish Patriots, however, to exist in a cultural and political formation which positioned them on the interface between antagonistic categories, rendering them radically insecure, simultaneously enabled and disabled, both coloniser and colonised. In terms of decolonisation, the one lasting achievement of Walker and like-minded Patriot critics was to broach the possibility of a resistance founded on something other than equality with England.[13]

Anglo-Ireland under Pressure

Ireland and Great Britain were legislatively linked by the Act of Union of 1800 in reaction to the attempted rebellion by the United Irishmen in 1798. In the following years the institutions and practices which had represented the island's identity as a separate though loyal kingdom disappeared one by one in a deliberate colonialist policy of cultural and political assimilation (Brown, 1972; McCormack, 1985; Vance, 1990). During this time, England,

Gaelic Ireland and Anglo-Ireland continued to vie for cultural, political and moral leadership of the nation and the right to employ the national nomenclature in the course of their own self- and other-constituting discourses.

Cultural nationalism was still dominated by Anglo-Ireland. Under pressure from an increasingly vocal native Irish element, however, this community was being forced to confront head-on the contradictions implicit in its own composite, not to say 'schizophrenic' (Leerssen, 1996b: 48) identity. Increasingly alienated from each of their constitutive identities, the Anglo-Irish attempted to negotiate various versions of 'transformist' and 'expansive' hegemony *vis-à-vis* their English and Irish competitors.[14] One important figure engaged in these cultural nationalist and hegemonic activities was the poet and critic Samuel Ferguson. Fiercely Irish, Protestant, nationalistic and loyal, Ferguson embodies the confusion which attended Anglo-Irish identity in the period after the Union. He belongs to what Roy Foster describes as a tradition of politically conservative Protestantism 'who formed a potential constituency for moderate nationalism' later in the century, a tradition in which 'Allegiances could be curiously confused, "interests" hard to define, the circle often squared' (1993: 62). Ferguson's criticism of James Hardiman's *Irish Minstrelsy* (published in the *Dublin University Magazine* in 1834) represents a comprehensive statement of Anglo-Irishness while also constituting one of the most skilful attempts to square the circle of modern (Anglo-) Irish critical history.

To appreciate the subtlety of Ferguson's critique of Hardiman it is necessary to place it in terms of the ongoing narrative (or rather narratives – competing and often hostile) of modern Anglo-Irish history. As remarked above, the Anglo-Irish position was shot through with contradictions on many levels, all stemming from its dual identity and its occupation of two coterminous discursive locations: a dominant one *vis-à-vis* the native Irish, a subordinate one *vis-à-vis* England. Depending on the subject's position, therefore, the Union could be seen either as a recommitment on the part of the English colonial masters to continued Anglo-Irish rule in Ireland or a tacit acknowledgement of the lack of success of that rule. For the descendants of those Protestant Anglo-Irish Patriots who had agitated for greater autonomy throughout the eighteenth century, the Union was an unacceptable (and illegal) breach of Irish sovereignty which thoroughly undermined the claim of the Anglo-Irish to lead the separate kingdom. For those shocked by the consequences of Paris in 1789 and Wexford in 1798 the Union offered a guarantee against the beast of revolution, whatever its form.

In this analysis, the Union was a necessary action which underpinned rather than undermined Anglo-Irish domination.

Ferguson operated on the interface between these two narratives. As he began his public and literary career in the early 1830s Anglo-Ireland was undergoing its most severe test since the Union itself, with the Catholic Emancipation Bill of 1829 and the Reform Bill of 1832 threatening to deprive the Protestant minority of the historical 'consensus' on which its dual identity relied. Ferguson was, famously, 'an Irishman and a Protestant', but more importantly 'an Irishman before I was a Protestant' (O'Driscoll, 1976: 15). As a Protestant, he was alive to what he saw as Catholic Irish ignorance and the likely consequences of popular rule such as had been threatened in 1798. This danger was again being threatened by O'Connell and, of all people, Hardiman. As Ferguson understood matters, in his book of Irish ballads Hardiman has attempted to appropriate the quality of Irishness to construct an exclusivist, antagonistic position which would deny Anglo-Ireland a role in the national history. At the same time, Ferguson is aware of the reasons for such a manoeuvre, and is both emotionally and intellectually attracted by the historical and aesthetic arguments underpinning it.

In the face of this dilemma Ferguson attempted in the course of his critique of Hardiman to construct a position for the Anglo-Irish in which they would be fully integrated into the national community and, in fact, the natural leaders of that community by virtue of their access to metropolitan reason in the form of a rational Protestant religion. In other words he made use of whatever was to hand – Irish history, English reason – to construct a powerful position for the national fraction he represented which would maintain their former power in spite of Catholic Irish agitation and English reform. Within the framework employed in this study, Ferguson's discourse represents a form of 'liberal' decolonisation in that, despite the rhetoric of difference and the frequently belligerent tone, it constitutes an attempt to insert the Anglo-Irish as a peculiarly privileged fraction of the nation while at the same time integrating both native and settler into the discursive economy of the 'mother' country.

This tactic may be observed in the comments at the end of his lengthy critique of Hardiman on the difficulties of translating Irish verse into English. The critique comes in the form of a four-part polemical essay in which Ferguson deals provocatively in turn with songs of love and comedy, melancholy and pathos, and the 'natural piety' characteristic of Irish people. In Part Four, Ferguson attacks and attempts to refute Hardiman 'in the various characters of antiquary, herald, historian, patriot, scholar, and pacificator' (1834: 517), and in an extended comic finale saves Hardiman's life when the latter is in danger of choking on his own words.[15] Thus, the bulk of the text is an open polemical attack on the political position Ferguson perceived underpinning Hardiman's volume. However, it is with the 300-word *coda* inserted at the end as a technical addendum to the open critique that Ferguson attempts to insinuate the Anglo-Irish into a leading role in national discourse:

> The main difficulty, and one which is in some cases insurmountable, consists in the multitude of words in the original forming a measure which frequently does not afford room for more than half the English expressions requisite for their adequate translation. This arises from the ellipsis of aspirated consonants and concurrent vowels, which frequently slurs three or four words into a single dactyl, and compresses the meanings into so small bounds, that the translator is driven either to lengthen the measure, and thus make his version incompatible with the tune of the original, if a song, and indeed with its spirit and character in any case, or else to double each stanza, and by a dilation as prejudicial to the genius of his subject as the over compression of too strict adherence, to lose the raciness of translation in the effete expansion of a paraphrase. (529)

It is immediately clear that the critic is operating in a different register in this passage, with the employment of a technical vocabulary signalling his complete command of the subject matter. As with Preston and Walker, this is significant in itself, demonstrating as it does both *Anglo*-Irish power when confronted with Irish history and Anglo-*Irish* validity in the face of English ignorance. Just as important in this instance, however, is the argument itself. In this passage Ferguson describes the danger of an Irish/English cultural confrontation without the mediating hand of the Anglo-Irish translator. Irish poetry has inherent value with its plenitude and complexity, its spirit and character and raciness. What it risks, however, in a straightforward encounter with English cultural discourse is the silence to which it had been condemned during the eighteenth century, or the misrepresentation which had ever been its fate at the hands of unsympathetic English translators. At the same time, the English language alone is seen as inadequate to the task of representing or containing this valuable experience; as long as it remains outside English discursive control, Irish cultural experience (in this case poetry and song) remains a real, though temporarily dormant, threat. Thus, a translator with access to both linguistic domains is necessary to resolve the misunderstanding and contempt on the one side, and the dissension and rancour on the other. Transposed into political terms, an inadequate translation is any English/Irish solution which ignores the Anglo-Irish impact on modern history; an adequate translation, on the other hand, will see the Anglo-Irish offering themselves as a sort of buffer between Irish demands and English expectations, with their own internal dominion confirmed and consolidated by their new central role.[16]

In this way, Ferguson constructed a new position for the Anglo-Irish which assured them a crucial role in the history of colonial relations between Ireland and England. As an Irishman he defends a discourse – Irish literature in the English language – with which Ireland can begin to assert its cultural validity; as an Anglo-Irish Protestant, he is on hand to lead this new Ireland, for without him Gaelic Ireland will continue to suffer the silence which marked its former confrontations with the modern world. Ferguson thus uses his critical discourse to validate Anglo-Irish identity and, in a reciprocal motion, achieves critical closure in terms of his political agenda. Nation (politics) and literature (culture) are explicitly linked; however, it is the unacknowledged relationship – that which exists between the *critical text* and Ferguson's political agenda – which in fact creates and enables the ideology of a 'natural' culture/nation nexus.

Like his contemporary Ferguson, Thomas Davis was an Anglo-Irish Protestant lawyer who studied at Trinity College, Dublin, and confined his interventions in the nationalist debate to creative and critical writing. Yet unlike Ferguson, Davis became an integral part of post-revolutionary nationalist mythology.[17] Like Tone, Emmet and Pearse, Davis was elevated into the nationalist pantheon where his work became valuable nationalist currency. How may we account for the difference in success between Ferguson and Davis?

Irish political resistance to English domination in the first half of the nineteenth century centred around the charismatic figure of Daniel O'Connell. At once 'Liberator' and 'King of the Beggars', O'Connell managed to harness that large fraction of the population which had been forcibly silenced throughout the eighteenth century and convert them into a huge moral-force movement. After the 'victory' of Catholic Emancipation in 1829, however, the limits of such resistance, as well as O'Connell's failure to attract Anglo-Ireland to the idea of Repeal, were exposed at Clontarf in 1843. It was left to 'Young Ireland', most notably the young lawyer Davis, to pick up the pieces after Clontarf and try to reshape them into an inclusivist, populist, nationalist ideology. The primary location of this activity was in the pages of *The Nation* where Davis and others set about this task of reinventing the culture of the people-nation through such devices as the ballad, polemical journalism and, of course, literary criticism.[18] As another example of the 'coloniser who refuses', however, Davis was beset by the same problems as Walker sixty years before, problems which he attempted, as did his predecessor, to shore up with a critical discourse.

The process whereby so many young, educated, middle-class Irishmen engaged in political activity in the 1840s has been discussed at length by Jacqueline Hill (1980). Evoking the concept of the 'scientific state', Hill argues that rationalism and utilitarianism were beginning to dominate English governmental policy at this time, and that the traditional order was being systematically dismantled to make way for the bureaucratic institutions of the modern British political machine. The Irish intelligentsia, unlike their British counterparts, were unable to find a role in this new order. Their old influence being eroded, they fell back on an ideology which guaranteed their centrality as guardians and articulators of the organic society – that is, the ideology of nationalism. Yet, even as these young lawyers, doctors and clerics were alienated from the scientific state, they were attracted by its appeal to education and reason and its promise of a fully functional society. Davis, for example, felt the tension between the nationalism which he embraced in reaction to his alienation from modern England, and his excitement about the possibilities of a state run on rationalist principles. While the scientific state was inadequate to the task of expressing the traditional nation as it was then being constructed by alienated writers all over Europe, many of these writers felt a sneaking admiration for the wonders of the modern world and suspected that the organic society would have to incorporate some of the new developments if it was to be successful.

This tension surfaces in Davis's work as a debate between the value of the traditional society and the benefits of modern education. In an essay 'On Study', for example, he recommends the pleasures of reading to young Irish people, but at the same time warns of the dangers which such reading might entail when not grounded in the safe, organic soil of the traditional national community:

> He must analyse as well as enjoy. He must consider the elements as well as the argument of a book ... Doing this deliberately is an evil to the mind whether the subject be nature or books. The evil is not because the act is one of analysis...The evil of deliberate criticism is that it generates scepticism ... But an earnest man living and loving vigorously is in little danger of this condition, nor does it last with any man of strong character ... Happiest is he who judges and knows books, and nature, and men (himself included), spontaneously or from early training – whose

> feelings are assessors with his intellects, and who is thoroughly in earnest. (in Anonymous, 1945: 90)

The critical function here represents the modern scientific state and the appeal to intellect and rationalism. Yet when it is 'deliberate', imposed, severed from instinct and feeling, it is an evil, as is England's attempted imposition of impersonal materialistic structures on Irish life. Only when it 'spontaneously' occurs can analysis and criticism be valuable, and this natural spontaneity can only be generated by participation in the national community. In an argument which curiously anticipates Yeats, Davis insists that the nationalist reader must consider the form (the elements) as well as the content (the argument) of literature; like Yeats, he is trying to ward off the accusation that nationalism, by necessity of its appeal to the emotions and the instinct, is unsophisticated and incapable of addressing the complexities of the modern world. His solution is to incorporate both regimes, insisting on (rational) analysis, but only as this is mediated by (national) feeling.

Given this contradictory structure, Davis's critical discourse was bound to show signs of strain. In the 1840s, the appeal to cultural nationalism was a 'radical' gesture which Davis has to develop in the face of English arrogance and O'Connellite derision: 'That a country is without national poetry proves its hopeless dulness or its utter provincialism. National poetry is the very flowering of the soul – the greatest evidence of its health, the greatest excellence of its beauty' (103–4). Critical formulations such as this opened up the national struggle on a whole new front, refuting the charge of 'provincialism' and placing the nation at the centre of human experience. However, beyond this expansive insight, Davis finds that an effective cultural nationalism will at a certain point be constrained by the nation itself and its overriding populist ideology. Thus, the truly great national artist will be one who knows and loves Ireland and all things Irish thoroughly and who is able to convey effectively this knowledge and love so that the reader may experience the national truth *within* the text itself. In other words, Davis's critical discourse is realist and historicist, relying on a crude mimeticism which completely undermines the original anti-provincial gesture. The will-to-criticise, which is meant to announce the modern validity of nationalism, is completely subsumed under the national truth which seeps out of the text and is available for appreciation by any 'earnest' nationalist without the need for cosmopolitan analysis.

When reviewing Carleton's *Tales and Sketches of the Irish Peasantry*, for example, Davis claims that the author 'is the historian of the peasantry rather than a dramatist' (110). The creative faculty is denied in favour of a mimetic faculty which serves larger national ends and sacrifices insight into the 'elements' as opposed to the 'argument' of the work:

> Born and bred among the people – full of their animal vehemence – skilled in their sports – as credulous and headstrong in boyhood, and as fitful and varied in manhood, as the wildest – he had felt with them and must ever sympathise with them. Endowed with the highest dramatic genius, he has represented their love and generosity, their wrath and negligence, their crimes and virtues, as a hearty peasant – not a note-taking critic. (111)

Carleton is to be valued, then, not for his critical capacity, but for his complete immersion in the fabric of national life and his ability to convey this as realistically and as fully as possible. The tone of anti-intellectualism in this passage is indicative of the direction in which dominant nationalist discourse was to go after Young Ireland. Davis's critical discourse signals a willingness to confront the modern world and the demands of the scientific state; immediately, however, the nation itself intervenes and tactically undermines any gesture which looks to signify beyond the national purview.

Davis was a radical decolonising intellectual confronting head-on the difficulties of constructing Irish identity in the terms made available by the colonial power. One means of warding off the implications of this contradiction was through the construction of a critical ideology predicated on the interdependence of culture and geography. After Davis, it would be very difficult for anyone to discuss Irish literature or Irish nationalism without being aware of the 'common sense' linking these two seemingly symbiotic categories. The location of this ideological operation was Davis's own discourse in which he employed the characteristic structures and codes of criticism to consolidate the ideology of a national literary tradition.

Irish Politics, English Culture

After Young Ireland, cultural nationalism became a central issue for Irish decolonisation. Once criticism, inspired by Davis and the scholar-critics of the late nineteenth century, created a space for a national culture, that culture did indeed begin to appear, celebrating what it considered to be distinctly Irish manners and experiences in distinctly Irish forms and styles. Cultural nationalism in Ireland in the second half of the nineteenth century demonstrates the need of decolonising formations to receive the sanction of history in their drive towards authenticity in the present. Such formations, as Anthony D. Smith argues,

> are bent on creating a new kind of individual in a new kind of society, the culturally distinct ethnic nation. This means returning to an idealized image of 'what we were', which will serve as an exemplar and guide for the nation-to-be. By returning to an ethnic past the community will discover a cognitive framework, a map and location for its unfocused aspirations. Similarly, 'our past' will teach the present generation not only the virtues of their ancestors but also their immediate duties. It will disclose to the community its true nature, its authentic experience and hidden destiny ... [In] all these movements intellectuals and intelligentsia play an important role. (1991: 140)

At the same time, Irish criticism after Davis provides an exemplary instance of the derivativeness besetting all discourses of resistance envisaged in oppositionalist terms, for the paradox of Irish cultural nationalism is that both critical and artistic discourses were thoroughly enmeshed with that which they ostensibly rejected. As David Lloyd explains:

> [The] desire to produce a canonical, 'major' literature and the subsequent adoption of a specific matrix of concepts through which the nature of the canonical is to be defined bring Irish nationalist theoreticians of culture into line with a conception of the canon and, more generally, of aesthetic culture that is ultimately linked to the legitimation of imperialism and to that mode of internal political hegemony, liberal democracy, which corresponds within the nation-state to global imperial domination. (1987: 3–4)

At the precise moment when Irish 'theoreticians of culture' (that is, critics) announced their opposition to metropolitan discourse by self-consciously linking culture with nation, they simultaneously confirmed the configuration of power and knowledge which constructed the terms of the colonial opposition in the first place. In deconstructive terms, we might say that the paradox of subordination is that it is parasitic on that which dominates it, relying for its identity on its consciousness as an oppositional discourse. Nowhere can the contradictions attending cultural nationalism in both its liberal and radical modes be discerned more clearly than in the work of one of the most self-conscious and paradoxical of modern Irish critics, W.B. Yeats.

Yeats's influence remains seminal in modern Irish cultural politics. Although he continued to refine and develop his position on literature, criticism and the nation throughout a long career, his early pronouncements almost single-handedly re-invented, and for a while dominated, the discourse of cultural nationalism in the post-Parnell era. As self-appointed spokesman for the dwindling Anglo-Irish Protestant population of Ireland, Yeats's task (following Ferguson's example) was to invent a history and an identity which would guarantee Anglo-Irish inclusion in, if not domination of, a restored Irish nation.[19] In pursuit of this overall strategy, his tactics included theosophy, mysticism, his own creative writing and, most importantly, a critical discourse which would simultaneously *reveal* and *embody* the essential validity of the Anglo-Irish tradition. In a series of lectures and articles throughout the 1890s and 1900s Yeats delineated a strategic position which allowed for the emergence of a practical literary and critical tradition. That is, he hoped to maintain the centrality of Anglo-Ireland to the emerging modern Irish nation by engaging in a critical discourse through which he could define the relationship between culture and nation, and thus shape what could legitimately be said about each.

Yeats described the model of literary history which would allow such a development in an article first published in the *United Ireland* journal in May 1893 (Yeats, 1988). There are, he claims, three stages in both literary and national development which correspond to the three stages of human growth: childhood, maturity, old age. The first stage is 'the period of narrative poetry, the epic or ballad period' (86), such as that represented in the works of ancient Greece and Rome. The second stage brought a 'dramatic period' (87) such as that experienced during Elizabethan England, while the last period brought the lyric which was the dominant form of nineteenth-century Romantic literature:

> The epic and the dramatic periods tend to be national because people understand character and incident best when embodied in life they understand and set among the scenery they know of ... But the lyric age ... becomes as it advances towards an ever complete lyricism, more and more cosmopolitan; for the great passions know nothing of boundaries. (89–90)

Ireland was re-entering its epic age at the end of the nineteenth century, with the philologists and scholars making available an entire new formation of texts on which to build a new legendary literature. For Yeats, this period and this literature would bring an energy and an enthusiasm necessary to revitalise an exhausted contemporary world, while undermining the decadence of European (specifically English) life and art. It is this anti-materialist strain that constitutes Yeats's most important contribution to contemporary Irish cultural politics.

So, literature and the nation are firmly enmeshed. As Yeats's friend Lionel Johnson wrote in an essay entitled 'Poetry and Patriotism' a year later: 'Poetry and patriotism are each other's guardian angels, and therefore inseparable' (1988: 101). Nationalism, however, also brought *nationalist*, as opposed to *national*, poetry, and this, for Yeats, Johnson and like-minded critics, brought a problem – albeit a problem that had quite useful implications for Anglo-Ireland. For although Davis and the poet-patriots of Young Ireland had successfully identified and tapped into Ireland's national rebirth during the nineteenth century, in so doing they had encouraged a literature that was populist, unsophisticated and unfit to represent the nation, newly imagined as a land of legends and heroes, of simplicity and spirituality.

Here, Irish literature could learn a thing or two from its English and European neighbours who for all their decadence were not unwilling to work at their art to make it a thing of beauty in itself as well as a representative of the beautiful nation. Instead of being national and liberating, post-Young Ireland literature and criticism had become provincial and coercive. As Johnson remarked, 'art' became an accusation of Englishness rather than an accomplishment of Irishness (93). This was the very quality which Anglo-Ireland with its education and its greater access to continental thought could bring (or in Johnson's argument, bring back) to Irish literature.[20] Yeats wrote in another essay in 1890: 'There is no great literature without nationality';[21] hence the need for Irish writers to write on national themes and to employ all the resources of national myth and language. But, he goes on, there can be 'no great nationality without literature', where 'literature' signifies a specific realm of activity policed by the artist-critics of Anglo-Ireland, entry to which is subject to their criteria. In other words, Ireland will never be a nation without a national art, and it will never have a national art without Anglo-Irish knowledge and experience.

Yeats continued to expand on this theme for the rest of his career. In so doing he not only provided an attractive defence of Anglo-Ireland, but he adumbrated that ideology formally in the course of his critical discourse. Vinod Sena has shown how Yeats moulded his criticism so that it becomes a form of autobiography, art and life mirroring and creating discursive spaces for each other. This is achieved by the use of anecdote, an interrogative style and the constant evocation of 'Yeats the poet' by 'Yeats the critic'.[22] As Seamus Heaney has commented: 'He merged his biographical self into the representative figure of the poet/seer, and he spoke as one both empowered and responsible beyond the limits of a private self' (1991: 788). The effect of this is to place creative discourse at the centre of contemporary Irish cultural experience while at the same time showing the experience out of which that culture has emerged. Yeatsian creativity and criticism engage, therefore, in a self-fulfilling prophesy in which art is grounded in life while life takes its impetus from art. Anglo-Ireland is doubly validated, therefore, finding its legitimacy in an art impregnated with contemporary experience and in a criticism founded on 'authentic' national art.

Yeats goes on to differentiate formally between the creative and the critical processes in overtly political terms. Replying to a comment that a national drama must be judged on national criteria, Yeats wrote that 'the true ambition is to make criticism as international, and literature as national, as possible' (Sena, 1980: 8). The paradox underpinning this statement is the paradox of liberal decolonisation: authentic national experience must be possessed of a 'supplementarity' which will secure for the Anglo-Irish a role in contemporary Ireland. In political terms this supplementarity is conceived as a necessary cosmopolitan dimension which will avert the evils of provincialism and insularity; in aesthetic terms it means that, along with style, technique and true art (as opposed to, for example, Davisite propaganda), *criticism* is enlisted as an element of authentic national experience. The fundamental contradiction at the heart of this notion, however, is that while insisting on the one hand that *his* life and *his* art should be commensurate, on the other hand Yeats denies the authenticity of a criticism which springs from its own object, or an art which evokes its own criticism. Criticism must vindicate Anglo-Irish experience by having a relationship with its cultural object which is at once congruous and discrepant.

This contradiction at the heart of Yeats's critical discourse did not prevent his ideas penetrating contemporary Irish cultural discourse and producing an easily accessible position from which Anglo-Irish subjects could attempt to avert their fatal decline, and which Irish-Ireland could appropriate or decry, according to its needs. Like Preston and Ferguson, however, Yeats's attempt to discover/construct a valid culture to equal metropolitan culture and thus form the basis for an equal political relationship was disabled at its conceptual moment, because the drive to assert equality in fact reinforced the structure of inequality. Liberal decolonising strategy of this kind was always going to be of only limited use in the construction of an Irish identity and an Irish history, because the terms in which that identity and that history could be articulated were thoroughly informed by metropolitan values and the logic of supplementarity through which English colonialist discourse functioned.[23]

Yeats's critical discourse emerged in the immediate post-Parnell period in dialogue with Irish Catholic intellectuals who for the first time were offering themselves as serious competitors for the cultural leadership of the nation. In a series of skirmishes, of which the *Playboy* riots of 1907 are the most important and famous, these intellectuals engaged with the rules which had once guaranteed Anglo-Irish domination of cultural nationalism to articulate a position founded on conservative Catholicism. Although this position owed more to early Yeatsianism than it liked to admit and had a broad enough base to accommodate various liberal and latitudinarian (indeed, as we shall see in the next section, *hybrid*) inflections, it was in fact its fundamentalist right wing that came to dominate it and the new Free State in the post-revolutionary period.

Ironically, one of the major tactics of this bid for leadership grew out of the thinking of an Anglo-Irish Protestant who, perhaps sensing the limitations of Yeatsian nationalism, attempted to secure a future for his section by creating a myth of the ideal nation founded on linguistic purity and emotional attachment. In his lecture on 'The Necessity for De-Anglicising Ireland' delivered before the Irish National Literary Society in Dublin on 25 November 1892, Douglas Hyde set out the terms for inclusion in the new nation:

> In conclusion, I would earnestly appeal to everyone, whether Unionist or Nationalist, who wishes to see the Irish nation produce its best – and surely whatever our politics are we all wish that – to set his face against this constant running to England for our books, literature, music, games, fashions, and ideas. I appeal to everyone whatever his politics – for this is no political matter – to do his best to help the Irish race to develop in future upon Irish lines, even at the risk of encouraging national aspirations, because upon Irish lines alone can the Irish race once more become what it was of yore – one of the most original, artistic, literary, and charming peoples of Europe.[24]

On this theoretical base an entire cultural superstructure becomes possible, the major planks of which are continuity of tradition and anti-metropolitanism. Along with the old Young-Irelander Sir Charles Gavan Duffy, Hyde continued to push a consistent de-Anglicisation line throughout the 1890s and 1900s, with the foundation in 1893 of the Gaelic League, probably his most important and enduring contribution. But this ideology also allowed for the development of a new critical discourse founded on the rigid demarcation of authentic national experience. An early example of this criticism can be found in Hyde's own *A Literary History of Ireland* published in 1899, where he writes:

> The present volume has been styled ... a 'literary History of Ireland', but a 'Literary History of Irish-Ireland' would be a more correct title, for I have abstained altogether from any analysis or even mention of the works of Anglicised Irishmen of the last two centuries. Their books, as those of Farquhar, of Swift, of Goldsmith, of Burke, find, and have always found, their true and natural place in every history of ENGLISH literature that has been written, whether by Englishmen themselves or by foreigners ... (1967: xxxiii, original emphasis).

The charge against these particular Irishmen is their infection by Englishness and their lack of emotional attachment to the land of their birth. In Hyde's discourse one need only identify with the tradition to become a part of it; thus, the Anglo-Irish, who would seem to be disqualified by their lateness and their racial and religious connections, can claim the national inheritance as a legitimate section of the new Ireland. This line of thought, stemming from late eighteenth-century Patriotism and consolidated by Ferguson in the 1830s (Leerssen, 1996b: 27–32, 181–4), was still an attractive alternative to Yeatsian nationalism for some Anglo-Irish subjects at the turn of the century.

Unfortunately for Hyde, however, having once initiated this discourse he finds it hijacked, its rules seized and redeployed, by those wishing to employ his anti-English tactic and to take the logic of Irish-Ireland to the extreme (Cairns and Richards, 1988b). In the first extract above he tries to forward the notion of a non-political movement where Unionist and Nationalist, Anglo-Irish and Irish-Irish, can share a platform of linguistic revivalism, anti-materialism and cultural exclusivism. As the new century opened, however, the scope of Irishness was narrowing all the time until after the institution of the Free State the critic Daniel Corkery (1967: 19) was able to define the nation in terms of three exclusive components: nationalism, land and religion.[25] Now, Anglo-Irish critics, whether inclined more towards Yeats or Hyde, were obviously concerned with the question of nationalism. By the same token, they could engage with debates concerning the island's rural heritage; the signification of the peasant did in fact become one of the major hegemonic engagements in the opening years of the century, climaxing once again in the *Playboy* riots of 1907. However, once Catholic Ireland had moved into a position of dominance in the cultural nationalist debate, the Anglo-Irish had little possibility of engagement in terms of the category of religion, and the choice effectively became one between deracination and marginalisation.

In retrospect, the notion of a de-politicised cultural discourse appears hopelessly naive. Hyde's implicit message was that literature and the nation were intimately connected in an inductive and deductive bind where every poem contained the essence of the nation, which was in turn made possible by the language and literature in which it could be imagined. This was put most succinctly by Patrick Pearse in an essay on 'The Spiritual Nation':

> The spiritual thing which is the essential thing in nationality would seem to reside chiefly in language (if by language we understand literature and folklore as well as sounds and idioms), and to be preserved chiefly by language; but it reveals itself in all the arts, all the institutions, all the inner life, all the actions and goings forth of the nation.[26]

Here, as with Hyde, as indeed with Yeats, language and literature are the 'essential' embodiment of the nation. The first modern Irishman to perceive this, according to Pearse, was Thomas Davis. Pearse had enough 'aesthetic' appreciation to elevate Mangan and Ferguson above Davis the poet, but it is still Davis who is the most Irish, who is set up as the direct precursor of the Gaelic League and who embodies the great truth that 'a nationality is a spirituality'.[27] It is to Davis, therefore, that the true nationalist writer should look for an authentic aesthetic model, and it is Davis who becomes the measure of Irish-Irish cultural politics. Fortunately dead and thus unable to contest the matter, Davis's religious and ethnic background is glossed over in Irish nationalism's drive towards its *moment of arrival* – that moment, as described by Chatterjee, when 'nationalist thought ... is not only conducted in a single, consistent, unambiguous voice, it also succeeds in glossing over all earlier contradictions, divergences and differences and incorporating within the body of a unified discourse every aspect and stage in the history of its formation' (1986: 51).[28]

The implications of Irish-Ireland's contestation of the rules of engagement may be seen in the work of many reviewers and critics in the first decades of the twentieth century, those 'radical' decolonisers concerned to adopt and adapt Hyde's programme for overtly political ends: the de-Anglicisation and concomitant Gaelicisation (as opposed to the Yeatsian 'Celticisation') of Ireland; the emphasis on a land-based economy and an anti-material rural culture; a national and personal identity founded on Catholicism and familism. This ideology finds its ultimate formulation in the two major critical works of the chief apologist for the new Free State, Daniel Corkery: *The Hidden Ireland* (1924) and *Synge and Anglo-Irish Literature* (1929).

Challenges to Cultural Nationalism

Despite different emphases, then, the dominant modes of Irish criticism – Anglo- and Irish-, radical and liberal – construed a definite and vital relationship between literature and nation. This did not exhaust the issue, however, as there have been other figures throughout modern Irish history who have worked to problematise the received ideology of a fatal link between cultural activity and national identity. Many of these figures had lived and worked abroad, as if the perspective of physical exile somehow enabled the development of a critical perspective at odds with narrow-gauge cultural nationalism. At the same time, as this latter discourse moved into a position of dominance in post-revolutionary Ireland it tended to exclude, marginalise or assimilate any critical vision which attempted to engage with cultural nationalism's terms of reference. Some, such as Oscar Wilde and Bernard Shaw, were disallowed access to the nation on account of their infection with metropolitan social values and intellectual habits. Others, such as James Connolly and Thomas MacDonagh, had their work remoulded so that it could be contained by the national narrative and made to signify in some positive way. And dominating the intellectual landscape was the threatening figure of James Joyce, whose work offered a profound challenge to those critical discourses – colonial or anti-colonial – founded on categories (such as narrative and authorship) and effects (such as representation and reference) which were ultimately derived from the Enlightenment's founding opposition between identity and otherness.

One alternative to standard decolonising practices is associated with the 'decadence' of Wilde and the aestheticist movement of contemporary Europe which finds its fullest anglophone formulation in his work.[29] According to Yeats, late nineteenth-century European civilisation was exhausted and the most obvious symbol of this was an art which existed purely for its own beauty, offering no apparent purchase on the real world. As the cultural revival proceeded, this myth becomes a double-edged weapon with which Anglo-Ireland, in as much as they rejected 'art for art's sake', could beat materialist

England, but which, in as much as they insisted on art above propaganda (their own vital national contribution), could be appropriated by Irish-Ireland for use against Anglo-Ireland itself. Both these adaptations are in fact distortions of Wildean aestheticism.

Wilde is especially interesting in this context for his refusal of the Arnoldian concept of the function of criticism which we can see, for example, animating the work of Yeats.[30] In 'The Critic as Artist' (1969: 99–224) he deliberately reverses Arnold's evaluative and functional constitution of the relationship between creativity and criticism, on the way exploding all those effects and qualities enabled by this relationship: sincerity, rationality, morality, creative precedence, critical distance. Most significantly, foregrounding the role of the critic is also a way of refusing narrow national bias:

> For, just as it is only by contact with the art of foreign nations that the art of a country gains that individual and separate life that we call nationality, so, by curious inversion, it is only by intensifying his own personality that the critic can interpret the personality and work of others, and the more strongly this personality enters into the interpretation the more real the interpretation becomes, the more satisfying, the more convincing and the more true. (162)

Truth is not found by collapsing the individual life into the national life but by recognising the protean quality of all human activity, and the importance of personality as opposed to community. The critical spirit, 'recognising no position as final', will allow us to 'annihilate race-prejudice by insisting upon the unity of the human mind in the variety of its forms' (219–20). In celebrating the 'free play' (219) of the mind Wilde has in fact produced an astonishingly *avant la lettre* vindication of what is known to modern cultural theory as 'intertextuality'.[31] At the same time, critical discourse should be, as the style and tone of 'The Critic as Artist' reveals, both a metaphor and an example of this.

Echoes of Wilde can be heard throughout the early twentieth century – in Synge's 'young man (who) will teach Ireland again that she is part of Europe'; in Moore's eschewal of his early commitment for a critical discourse entirely free of any vestige of nationalism; in Buck Mulligan's entreaty to Stephen Dedalus to help him 'Hellenise' the island (Joyce, 1993: 7); perhaps also in the middle-period Yeats who sought 'monuments of unaging intellect' in the lonely tower of his soul.[32] The very situation which allowed Wilde to adopt and develop this attitude – his Irish 'otherness' located at the heart of the metropolitan power – also denied him full access to a nomenclature dominated by cultural nationalism. And with his exclusion an alternative critical politics, based upon an alternative version of the relationship between culture and nation, was lost.

Another discourse subsequently marginalised from post-revolutionary Ireland was that of class which, in both revolutionary and evolutionary modes, was sacrificed for an overarching nationalist imperative (Brown, 1985: 102–37). In so far as he sought an accommodation with nationalism as outlined in his book *Labour in Irish History*, James Connolly could be absorbed into the nationalist canon. The 'socialist key' (1983: 135) which he used to open the door on what he considered the true cause of Irish revolutionary activity throughout history was conveniently misplaced in post-revolutionary Ireland. Connolly claimed that all superstructural activity, 'War, religion, race, language, political reform, patriotism – apart from whatever intrinsic merits they may possess – all serve in the hands of the possessing class as counter-irritants, which function by engendering heat in such parts of the body politic as are farthest removed from the seat of economic enquiry, and consequently of class consciousness on the part of the proletariat' (3–4). State-building, middle-class nationalism with its cultural nationalist ideology, was merely part of the current phase of an international capitalist narrative in which bourgeois and aristocratic classes battled for the means of production as a prelude to the dictatorship of the proletariat. For Connolly, therefore, the land struggle, the national struggle and the class struggle were all one, a belief which he realised in the Easter Rising of 1916.

Similarly, for the Fabian socialist Shaw, nationalism was a false consciousness which, if not incorporated into the larger issue of social justice, was a detriment to true Irish interests and should be jettisoned. In a section entitled 'The Curse of Nationalism' in the 'Prelude for Politicians' to his play *John Bull's Other Island* (1906), Shaw wrote:

> Nationalism stands between Ireland and the light of the world ... The great movements of the human spirit which sweep in waves over Europe are stopped on the Irish coast by the English guns of the Pigeon House Fort. Only a quaint little off-shoot of English pre-Raphaelism called the Gaelic movement has got a footing by using Nationalism as a stalking-horse, and popularizing itself as an attack on the native language of the Irish people, which is most fortunately also the native language of half the world, England included. (Laurence, 1971: 842)

For Shaw, the Gaelic movement and the cultural revival out of which it sprang were at best a symptom of, at worst a distraction from, Ireland's real problem, which was one of social justice.

Moreover, in an early example of the derivativeness line, Shaw maintains that Irish nationalism was a phenomenon inspired by English history and European *fin-de-siècle* idealism, and consequently shot through with contradictions. The English language, therefore, is Ireland's only hope of salvation rather than its doom. Such an argument was repugnant to dominant post-revolutionary Irishness and the image of a nation of Gaelic-speaking peasants.

For both Connolly and Shaw there are obvious, if implicit, implications for cultural and critical discourse. Culture does not represent the nation, the nation does not generate a culture, except in so far as nationalist critical discourse construes it so. As a discrete, autonomous, intellectual practice, criticism's function should be to understand and describe those particular local, national myths by the light of the larger, international, universal truth of social justice.

If these challenges from outwith the discourse of decolonisation have tended to be marginalised in postcolonial Ireland, certain discrepant notes from *within* cultural nationalism regarding the culture/nation nexus have also been silenced or ignored. The poet, revolutionary and critic Thomas MacDonagh, for example, was for many years seen 'to represent the perfect, tragic fusion of literature and politics in Ireland' (Storey, 1988: 8). When the Easter Rising came to represent (as it did in the revisionist historiography of the 1970s) fundamentalist nationalism (Lyons, 1973; O'Brien, 1972; Shaw, 1972), MacDonagh's literary critical discourse was by association implicated in the same narrow-gauge nationalism which characterised the work of Moran, Pearse and Corkery. But MacDonagh's work, as Luke Gibbons has argued (1991; 1996: 156), belies this simplistic depiction of nationalism. In the posthumous collection *Studies Irish and Anglo-Irish*, by transferring attention from a 'Celtic essence' to an 'Irish note', MacDonagh insists on the validity of Irish literature in the English language and looks for an accommodation with Anglo-Ireland. The issue is not one of essence at all, but of sincerity:

> We have now so well mastered this language of our adoption that we use it with a freshness and power that the English of these days rarely have ... The loss of (Gaelic) idiom and of literature is a disaster. But, on the other hand, the abandonment has broken a tradition of pedantry and barren conventions; and sincerity gains thereby ... Let us postulate continuity, but continuity in the true way. (1916: 169–70)

Here, the critic identifies and celebrates the 'freshness and power' of English as used in Ireland. Further, he applauds the disappearance of an effete and decadent late Gaelic tradition as itself a perversion of the 'true' Irish note. Reading from MacDonagh, later critics might use this notion to account for the emergence of powerful Irish writers of English such as Synge, Joyce and O'Casey, all excluded at one time or another from the nationalist canon, without slipping into the racialism which has dogged cultural nationalist debate since the eighteenth century. As an abstract linguistic tendency rather than an innate racial capacity, the 'Irish note' should be fully available to the Anglo-Irish in MacDonagh's model of cultural production. The subtlety of this position is unrecognisable as the vulgar fundamentalist nationalism attacked by revisionist discourse.

That a form of radical (bordering on fundamentalist) nationalism won out during the revolutionary period is generally accepted, however, even by Gibbons who is otherwise keen to stress the expansive vision of many figures on the margins who were subsequently written out of, or assimilated into, the dominant nationalist narrative. The Free State soon adopted the conservative and authoritarian structures which were to dominate it for the next half century (Brown, 1985: 45–78; Cairns and Richards, 1988a: 114–38). The national imperative, in as much as it came to be identified with certain religious, familial and cultural ideologies, dominated individual experience and produced a national identity founded on racial exclusivity, individual repression and fear of difference. The version of nationalism which came to dominate the Irish Free State, Éire and eventually the Republic of Ireland in the years after 1922 had a significant investment in cultural nationalism, founded on the ideology of a fully self-present national subject aware of her place in an ongoing national narrative and capable of intending and transmitting national meaning.

The irony whereby the critical tradition based on the interdependence of culture and nation received its most profound challenge in the same year the Free State was constituted has been frequently remarked. James Joyce's *Ulysses* rehearses the categorical discourses of cultural nationalism so that they emerge as quoted, radically relative and historically placeable phenomena, discourse *for* and *from* specific subjects in specific spatial and temporal locations, rather than the unique, essential, authentic experiences imagined in colonising and decolonising activity.[33] Joyce's work is parodic, as many have remarked, but it is a subversive parody in which he has seized the rules of dominant decolonising discourse and disrupted what he sees as its flawed identitarian message. The creative artist cannot escape the languages of liberal and radical decolonisation;

indeed, Joyce does not actually want to 'escape' them as he remains affiliated to a certain notion of Irishness which can only be imagined in terms of those discourses.[34] However, by exposing them *as* discourses, as formal and historical organisations of signs, he can displace them so as to ward off, at least temporarily, the disabling structures into which they are locked. And the most disabling structure which *Ulysses* exposes is Irish literary criticism and the twin assumptions upon which it rests: a natural link between culture and nation, and a natural aporia between primary (imaginative) and secondary (critical) discourses.

A variety of critical strategies were brought to bear on Joyce's work, during his own life time and after, in an effort to contain the threat to the 'natural' culture/nation nexus promulgated by cultural nationalism. In the course of these critical productions, Joyce's challenge is invariably lost, much of the time even by those looking to defend his work.

Initial Irish reactions to *Ulysses* were predictably unfavourable. The poet and critic Shane Leslie set the tone with a scathing piece in the *Dublin Review* of September 1922 (Deming, 1970a: 300–3), during the course of which he referred to the book as 'morbid and sickening', 'a Cuchulain of the sewer ... an Ossian of obscenity', 'the devil's book', 'a Sahara that is as dry as it is stinking', and its author as one who 'revolves and splutters hopelessly under the floor of his own vomit'! By linking what he understands as its anti-Catholicism ('In its reading lies not only the description but the commission of sin against the Holy Ghost'), its acknowledgement of female sexuality and desire ('a very horrible dissection of a very horrible woman's thoughts'), and its challenge to traditional literary and linguistic forms ('abandoning all reasoned sequence of thought and throwing the flash and flow of every discordant, flippant, allusive or crazy suggestiveness upon paper without grammar and generally without sense'), Leslie correctly identifies the nature of Joyce's challenge to the conception of Irishness which had become dominant in the years leading up to the revolution.

A generation later the book may have changed but the nature of the challenge is still cast and rejected in the same subjective, sexual and linguistic terms. In the *Saturday Review of Literature* in 1941 Oliver St John Gogarty compared Joyce unfavourably with Yeats, representing the matter as a 'choice between the Logos, the Divine Word, "this godlike Reason", and the large discourse and senseless mutterings of the subliminal mind's low delirium' (Deming, 1970b: 750–1). A *Dublin Magazine* reviewer of 1939 opined that with *Finnegans Wake* Joyce had 'undone the work of the Irish literary revival and given us something as monstrous and ugly as the later-day Abbey brogue' (Deming, 1970b: 710–12). Another of the 'living' characters from *Ulysses*, John Eglinton, reluctantly acknowledged Joyce's genius but warded off the implications of his challenge to established literary-critical practices with the opinion that 'he is not specially interested in "literature', not at all events a well-wisher ... his interest is in language and the mystery of words' (1935: 150).

The strongest and most coherent domestic attack on Joyce in the 1920s and 1930s came from another opponent of dominant nationalism, Seán O'Faoláin, who in a number of articles and reviews consistently decried the lack of 'reality' in Joyce's work and the fact that his art 'comes from nowhere, goes nowhere, is not part of life at all' (Deming, 1970b: 389–90). O'Faoláin saw Joyce's experiments with language and literature as 'a revolt against the despotism of fact' (Deming, 1970b: 395–6), thus recalling Matthew Arnold's Celticist discourse of half a century before, unaware, perhaps, of the implicit structure of power and knowledge on which this discourse relied (Smyth, 1996a). The realist aesthetic purveyed by O'Faoláin throughout his critical and creative work was part of an early revisionist analysis of the recent revolutionary period. An emphasis upon realism would shake post-revolutionary Ireland out of the essentialist fantasies it had adopted as part of its radical decolonising programme. But as *Ulysses* and *Work in Progress* (published as *Finnegans Wake* in 1939) demonstrated, the move towards realism was ultimately no less fantastic than nationalism's necessary fictions. Indeed, as we have seen in Chapter 1, for many critics, realism relies upon certain assumptions – linear time, self-regulating subjects, the transparency of language – which in fact reinforce the values which underpinned colonial discourse. When he complains that 'the half-aesthetic behind *Anna Livia Plurabelle* denies that art can add a jot to reality' (Deming, 1970b: 399–400), O'Faoláin misses the complexity and completeness of Joyce's attack on a reality limited by a particular critical model of the relations between the cultural and political spheres.[35]

Although the overall trajectory of critical response was negative, Joyce did have Irish defenders in the years before 1948 – Mary and Padraic Colum, Yeats himself, Joseph Hone, and others. There was always a tension between reclaiming Joyce because of his obvious Irish connections and the kudos of 'owning' such an important world-famous artist, and disclaiming any similarity between his version of the national identity – perceived even by many of those out of sympathy with dominant nationalism as salacious and corrupt –

and official Irishness. The fullest Irish defence of Joyce came from those who had contact with his Paris circle and who were sympathetic to his aesthetic and historical preoccupations. Two of these writers, Samuel Beckett and Thomas McGreevy, contributed to a volume of criticism on *Work in Progress* which announced its affiliation with Joyce's work (and indeed bears the marks of his editorial influence) with its playful title: *Our Exagmination Round His Factification For Incamination Of Work In Progress*. But this text was more than just an acknowledgement of Joyce's linguistic experiments through neologisms and wordplay; rather, it offers an early example of the particular challenge awaiting all Joyce criticism. In attempting to explicate a literary practice which refuses the concept of metadiscourse (such as literary criticism), the contributors to the collection were obliged to confront the paradox of a textual subject who can apprehend meaning in the primary text only through the mediation of a secondary text. Because of its materiality, its in-the-worldness, the text must/will be criticised; but at the same time there is something in the text which, by staging and re-presenting the notion of an extra-textual reality to which the text must answer, resists such an operation.[36] This paradox is addressed by Beckett in the volume's opening essay:

> The danger is in the neatness of identifications ... And now here am I, with my handful of abstractions, among which notably: a mountain, the coincidence of contraries, the inevitability of cyclic evolution, a system of Poetics, and the prospect of self-extension in the world of Mr Joyce's 'Work in Progress' ... Must we wring the neck of a certain system in order to stuff it into a contemporary pigeon-hole, or modify the dimensions of that pigeon-hole for the satisfaction of analogymongers? Literary criticism is not book-keeping. (1961: 1, 3–4)

The critic confronts the paradox of trying to address, even sympathetically, a text which refuses the boundaries – institutional and discursive – between different kinds of writing, between inside and outside the text, between imaginative *writing* and critical *righting*, between Joyce and Beckett. The intervention of the self at the moment of criticism is an attempt both to subvert the concept of a metadiscursive subject, and to point out the manner in which critical discourse actually *produces* what it purports to *discover* in a certain kind of writing it designates primary and imaginative. In this way Beckett looks to mirror Joyce's refusal to collapse the moment of writing into a pre-existent or exterior reality, demonstrating instead how 'literature operates in a realm that undermines alternatives and the logic of identity' (Attridge, 1992: 128). Beckett's essay is an example of what Paul Smith, explicating Derrida, calls a 'criticism under erasure' (1988: 43) – a criticism, that is, which attempts to negotiate the paradox of a necessary search for meaning (a search in which Joyce was complicit but ultimately no more authoritative than any other critic of his work) and a concomitant need to acknowledge the constitutive moment of discourse, both literary and critical.

Beckett's essay (and to a lesser extent McGreevy's[37]) attempted to meet the challenge set by Joyce, a challenge which, because of the terms in which it is framed, could only ever be partially successful. Joyce famously wanted 'to forge in the smithy of my soul the uncreated conscience of my race' (1977: 228). To that end, he announced his affiliation with a radical Irish subjectivity, both by making Ireland the *raison d'être* of his work and by subverting a (English) culture which functioned as a component of colonial domination. More than this, however, he had to resist his own affiliation to one or another mode of decolonisation which was already overwritten with the disabling languages of domination and subordination. Joyce's resistance to Irish decolonisation takes the form of a challenge to a tradition of literary criticism organised around the notion of an indissoluble link between national culture and nationalist politics. As another Irish writer who was to refuse the limitations of being Irish, Beckett remains one of the very few Irish critics to recognise and respond to this challenge.[38]

In this chapter I have attempted to trace the emergence of an Irish critical discourse predicated on the ideology of a symbiotic relationship between culture and nation, and to identify the function of this discourse as part of the strategies of liberal and radical decolonisation. In this last section I have delineated some of the challenges to that ideology as described by various subjects engaged in various critical practices. These challenges either offered variations on the received culture/nation nexus or refused that concept entirely in favour of alternative models of cultural production. In recent years Joyce has developed into an emblematic figure in both these narratives. In so far as he was an *Irish* writer, Joyce was a radical critic of English domination of Ireland; but in so far as he refused the limitations of both Irishness and Englishness, he managed to incorporate both these categories into a writing practice which itself refused incorporation into any dialectical discourse predicated on a subject who, like the critic, purports to stand outside his own constitution in discourse. After Joyce's revolution of the word and

Ireland's revolution of the sword, both the role of the national subject and the function of a critical discourse in which such a subject was constituted had to be renegotiated. The task of the second part of this book is to examine the ways in which such renegotiations were conducted in 'postcolonial' Ireland.

PART 2

LITERARY CRITICISM IN IRELAND 1948–1958

4 Criticism in the Thin Society

One reason for much of the resistance to 'postcolonialism' as an analytical category is that 'post' in this context refers both to temporal and psycho-cultural discourses.[1] A geo-political formation is officially 'post'-colonial after the withdrawal of the colonialist power. In Ireland, this process occurred piecemeal, albeit with three significant moments: the Anglo-Irish Treaty accepted by the Dáil in January 1922, de Valera's constitution of 1937 in which the Irish Free State became Éire, and the declaration of the Republic of Ireland and secession from the British Commonwealth in April 1949. Many critics and theorists, however, discern an essential continuity of colonialist and decolonising discourses into the 'postcolonial' period, a continuity which gives the lie to the assertion of a chronological narrative (pre, then post; before, then after) implicit in the term. Thus Fanon:

> Nationalism, that magnificent song that made the people rise against their oppressors, stops short, falters and dies away on the day that independence is proclaimed. Nationalism is not a political doctrine, nor a programme. If you really wish your country to avoid regression, or at best halts and uncertainties, a rapid step must be taken from national consciousness to political and social consciousness ... if nationalism is not made explicit, if it is not enriched and deepened by a very rapid transformation into a consciousness of social and political needs, in other words into humanism, it leads up a blind alley. (1967: 163, 165)

Glossed by Edward Said, this means 'that unless national consciousness at its moment of success was somehow changed into a social consciousness, the future would hold not liberation but an extension of imperialism' (1993: 323). In similar vein, Ashis Nandy describes colonialism 'as a shared culture which may not always begin with the establishment of alien rule in a society and end with the departure of the alien rulers from the colony' (1983: iii).

Which is to say: neither subject nor state possesses a switch that is simply flicked from 'colonial' to 'postcolonial' with the withdrawal of the 'alien rulers'. Understood by Fanon and Nandy (as we saw at length in Chapter 1), the reasons for the persistence of (de)colonialist discursive modes into the 'postcolonial' era are both materialist and psychological, involving the national bourgeoisie's continuing subservience to their former imperial masters, their desire to maintain the power structures inherited from the colonial polity rather than instigate new civil and political programmes, and the national subject's inability to imagine modes of thought beyond those which structured the narrative of decolonisation. All told, the 'post' in 'postcolonial' is a much more complex prospect than it may at first appear, and in many important respects this reflects the situation one finds in Ireland's first 'postcolonial' decades.

In the years after 1922, Pearse rather than (liberal) MacDonagh, (socialist) Connolly or (feminist) Sheehy-Skeffington emerged as first among equals. It was Pearse's version of the national narrative, emphasising Catholicism, ruralism and anti-modernism (for which read anti-Englishness), which came to be adopted in the new state (and which in time also came to provide the target for oppositional and revisionist critiques of 'nationalism'). Despite the best efforts of its liberal and left wings, radical decolonisation was commandeered by a nationalist bourgeois élite which tried to arrest the decolonisation process at the point where it assumed control of the state apparatus left vacant by the off-shore power. The new state was nearing its 'moment of arrival', that moment when the cacophony of Irish history would be orchestrated into 'a single, consistent, unambiguous voice' (Chatterjee, 1986: 51). Postcolonial Ireland, that is, came to a sense of itself as a hegemonic rather than counter-hegemonic force, and as David Lloyd argues, thus 'becomes an effective brake on the decolonising process culturally as well as economically' (1993: 7). In this way, the attempt to formalise the abstractions of decolonising discourse in national(ist) institutions and practices in the years after 1922 actually reinforced social and political hierarchies even as the state claimed to be the agent of liberation from such hierarchies.

Political freedom (partial though it was) proved a heavy burden for the pro-Treatyites in the 1920s, and this was to continue into the 1930s when politicians and intellectuals alike became preoccupied with the relationship between pre-revolutionary images of Ireland and the reality of living in an impoverished statelet on the edge of Europe, in the middle of the greatest economic crisis ever experienced in the West, on the brink of what promised to be the most terrible war the world had even seen. Fianna Fáil's long domination of Irish politics began when they were elected to power in 1932, at which time de Valera found himself in control of a state against which only ten years before he had promoted a civil war.

The isolationist policies of the 1930s, the 'rather opportunistic form of neutrality' (Kiberd, 1991: 1316) that set the tone for the South's foreign policy in the latter part of the century, and the general political and cultural exhaustion of the post-revolutionary period meant that the fledgling state emerged in the years after 1945 as a nation self-sufficient in little except rhetoric. Colonialism had not ended 'with the departure of the alien rulers from the colony'; for many, Ireland after England meant 'not liberation but an extension of imperialism'. The brave new nation imagined in decolonising discourse appeared to have wound up in something resembling that 'blind alley' identified by Fanon. To put it in the more specific, if no less figurative, terms of Roy Foster: 'the weight of the historical palimpsest continued to impose its constraining pattern on the pre-occupations of Irish politics' (1989: 577).

Although political and cultural critique had been implicit since 1922, the 1950s sees the first stirrings of a coherent and widespread revision of this 'constraining pattern'. By 1948, it seemed clear to many that despite the rhetoric of independence and freedom, modern Ireland had not fulfilled (or not been allowed to fulfil) the promises which had fuelled pre-revolutionary discourses of decolonisation. According to both contemporary accounts and subsequent historiography, in fact, the decade between 1948 and 1958 was a disappointing, not to say miserable, time in modern Irish history – economically, politically, socially and culturally.[2] This malaise manifested itself in a number of ways. First, the effects of the 'Emergency' (as the war years were referred to in official terminology) were felt well into the following decade, and the standard of living in the part of the country not affiliated to Great Britain continued to lag behind most of Europe despite the influx of millions of Marshall Aid dollars (Lee, 1989: 305; Lyons, 1973: 589). Neither Fianna Fáil nor two Coalitions, neither de Valera nor Costello nor MacBride in four rapid changes of government within ten years, managed to halt the economic decline and the social and cultural stagnation which accompanied it. Two decades of sterile debate on corporatism, vocationalism, state regulation and other economic philosophies had produced a situation which saw Ireland possessing, in the words of Joseph Lee, 'the lowest living standards, the highest emigration rates, the worst unemployment rates, and the most intellectually stultifying society in northern Europe' (1979: 24).

Economic under-achievement contributed to another of the major manifestations of the malaise: emigration. Always an important factor in the cultural and economic health of the island since the Famine, it reached crisis proportions in the 1950s with what Terence Brown calls 'a widespread rejection of the conditions of rural life' (1985: 183) augmenting the numbers who, given the emigration culture, would have been expected to leave the country anyway. Lyons reckons that emigration rates for the decade 1951–61 'were higher that for any comparable period in the twentieth century' (1973: 625). The predominant destination of these economic emigrants was England, and this, allied to the new Republic's continuing reliance on the vagaries of Britain's economic performance and the failure of the much-heralded Gaelic language revival, fuelled a demoralising psychology of postcolonial dependence which nearly thirty years of 'independence' had done little to combat.

With the rejection of rural life and the consistently poor economic performance came a crisis in the cultural identity of the state, weaned as it had been during the early years of independence on an ideology of racial authenticity, rural simplicity, agriculture as the favoured national industry and Catholic familism as the dominant social culture. The residue of these discourses persisted throughout the 1950s, however, and the sense of Irish identity which came to dominance during the revolutionary period continued to exert a powerful influence on contemporary Irish experience. 'In post-independence Ireland', as David Lloyd has commented, 'the historical "victor" has clearly been a politics predicated on the constitution of national identity' (1987: xii); any indications of a threat to that identity, such as Dr Noel Browne's 'Mother and Child Scheme' of 1951 – a gesture in the direction of Welfarism which many saw as the thin end of a socialist wedge – or the apparent acceptance, implicit in the declaration of the Republic in 1949, of the border with Northern Ireland as *de facto*, excited reactionary interventions such as those attempted by a powerful Catholic hierarchy and a regenerated IRA.

The effects of this crisis were felt right across the public and private sectors, and two highly critical documents which come at either end of the period in question – Professor Thomas Bodkin's *Report on the Arts in Ireland* (1949) and T.K. Whitaker's *Economic Development* (1958) – demonstrate the need felt by contemporary commentators to address in a responsible and realistic manner the politico-cultural paralysis which had apparently stricken the country. They also demonstrate, however, how limited such critiques were as long as they continued to work within the discursive parameters of dominant decolonisation, and how much Irish identity in this supposedly 'postcolonial' epoch still depended on its colonial history for a sense of its own reality.

Bodkin roundly criticised the lack of state support for the arts since 1922, describing and deploring the situation 'in which it has become justifiable to say of Ireland that no other country of Western Europe cared less, or gave less, for the cultivation of the Arts' (1949: 9). Such, argues Bodkin, had not always been the case in Ireland, and he cites Thomas Davis as an example of the patriot-statesman who recognised the centrality of Art to a modern Western society (7). In the course of a detailed argument Bodkin sets out a fifteen-point plan for the contribution of a rehabilitated Arts culture to modern Irish society. The author, however, can only imagine contemporary official philistinism and its potential solution in terms of the received languages of liberal and radical decolonisation – that is, in terms of Ireland's similarity to, and difference from, other comparable nation-states, especially England. 'The Irish Advisory Committee on Cultural Relations', for example, 'is, in effect, charged with a large part of the responsibility of doing in Ireland what the British Council was appointed to do for England. The contrast between the two bodies, constitutionally and financially, is extreme' (49). Although criticising what he perceives to be the unsatisfactory relationship between the cultural and political spheres, Bodkin's report stills functions within a coherent narrative of nation-statehood. That is to say: although offered as a critique of post-revolutionary Ireland, this 'critical' discourse operates within systems of thought which have their roots and their significance in decolonising (and colonialist) discourse.

Although emanating from a different discursive sphere and written a decade or so later, the same is largely true also of Whitaker's 'radical' (Lee, 1989: 347) report. This ambitious and well-educated civil servant pulled no punches in his description of contemporary Ireland, a society in which, as he wrote, the endemic mood of despondency and anxiety was in constant danger of degenerating into frustration and despair, and in which:

> After 35 years of native government people are asking whether we can achieve an acceptable degree of economic progress. The common talk amongst parents in the towns, as in rural Ireland, is of their children having to emigrate as soon as their education is completed in order to be sure of a reasonable livelihood. To the children themselves and to many already in employment the jobs available at home look unattractive by those obtainable in such variety and so readily elsewhere. (1958: 5)

Already, however, as adumbrated by the last sentence, Whitaker's critique is compromised by its engagement with the comparative terms of traditional decolonising discourse. Like so many other critics of post-revolutionary Ireland, Whitaker recognises the need 'to shut the door on the past' (9), only to discover that such a gesture is anticipated by an established structure of Irishness and non-Irishness received from the colonial period. *Economic Development* was indeed a 'radical' intervention, and on those terms it may be said to have been brave and necessary. At the same time, however, it was also subject to the limits of the radical framework from which it emerged, and as such, always going to be a limited gesture in the ongoing narrative of Irish decolonisation. This pattern of genuine critique undone by the limits of received decolonising discourse is one we shall discover recurring throughout Irish criticism of the 1950s.

Besides such official critiques, the principal location of the disillusion felt at the performance of postcolonial Ireland was the work of the writers and critics of the period. From Patrick Kavanagh's *The Great Hunger* (1942), through Seán O'Faoláin's editorials in *The Bell*, on to Myles na gCopaleen's increasingly scathing column in the *Irish Times* – in every form and at every opportunity Irish writers were weighing the achievement of what O'Faoláin referred to as 'the thin society' which was post-Treaty Ireland and generally finding it wanting (1949: 376). At the same time, many writers (including the above named) were caught between their alienation from, and their commitment to, the new Ireland (Deane, 1986: 210ff.). The impact of this ambivalent confrontation between writer and society is one of the dominant discursive features of the period, in 'primary' cultural activity certainly, but more significantly in those 'secondary' critical practices which looked to intervene self-consciously in the relationship between cultural and political identity.

A number of issues were impacting upon Irish critical discourse at this time, four of which I wish to mention briefly here before going on to explore them at greater length in the following chapters: censorship; cultural isolation; the development of international Irish Studies; professionalisation. Turning to censorship first, whatever the original intention in 1929, by the 1950s the state's laws had degenerated into witch-hunting and anti-intellectualism of the worst kind. The actual effect of these laws on the produce of individual writers is impossible to calculate, but the under-realised careers of many of the artists (and critics) of the period is perhaps one of the best indications of the general economic and cultural lassitude which appeared to have gripped the country in the 1950s.[3] Similarly, isolation meant that Irish critics found it difficult to remain abreast of international critical debates, and there was as a result little opportunity to break free from the methodological and conceptual debates already explored, sometimes to exhaustion, by previous generations of Irish critics. Although referring to a special case, W.J. McCormack's point, that at this time, 'The impact of *Scrutiny* and *Horizon*, even of T.S. Eliot's *Criterion*, is nowhere evident' (1995: 85) in Irish critical discourse holds good. On the other hand, as McCormack goes on to suggest, this situation may not have been the disaster it was felt to be by many contemporary (and subsequent) critics, as it meant that 'by default, the study of Irish literature was to work out its own procedures and methods' (85).[4]

These factors were accompanied by the consolidation of the 'Irish Literary Revival' as an established area of international literary critical inquiry. This period witnessed the rise of specialist publications, conferences and symposia, biographies, critical studies and general academic engagement with the island's modern literary heritage. It also saw the first work of many of the international scholars and critics who were to be influential in Irish Literary Studies over the next three decades – Richard Ellmann, A.N. Jeffares, Hugh Kenner, David Krause, William York Tindall, etc. – as well as the formation of a canon of Irish Literature which has remained more or less dominant to the end of the century, with Synge, Yeats, O'Casey, Joyce and Beckett at the centre, and a host of 'minor' figures at the margins. Like many of their fellows since, Irish critics of the 1950s were caught between resentment at this usurpation of their 'patch' and flattery that issues which were still of direct relevance to them were being paid such impressive intellectual attention (Arnold, 1991: 87–101). Often reduced to hackwork in the intellectually moribund atmosphere at home, many were happy to be drafted into these debates for their local knowledge and expertise. As a consequence, much of the domestic critical discourse of the period reveals a dialectical movement between scholarly pronouncement and anecdotal reportage.

Finally there is the issue of the growing professionalisation of critical discourse which was making itself felt during this period. Towards the end of the 1950s, under the impetus of new administrative blood and new political thinking, the Republic made a determined effort to plan its way out of the economic stagnation which had dogged it since independence. The famous names here are Lemass and Whitaker, but the important thing to note is the ideological change in temporal and spatial perspective which came to dominate political and cultural activity in the late- and post-de Valerean era – from insularity to a re-entry into the world of international affairs (joining the United Nations, suing to join the European Economic Community, and the move towards Keynesian economic philosophy), and from past-oriented social and cultural policy to a recognition of the need to invest in the nation's future. These changes, indicative of a general shift away from the national self-obsession which had dominated Irish life since 1922, are related to the calls which sounded throughout the immediate postwar decades for a move away from general humanist, holistic, intellectual activity towards more specialised and scholarly practices. This – whether understood as the *trahison des clercs* or as a necessary professionalisation of the intellectual faculty – was clearly another of the factors impinging on the question of the function of criticism in relation to the nation (O'Dowd, 1985, 1988).

In many significant ways, then, criticism's traditional role as a crucial site for hegemonic encounters between different ways of figuring the culture/politics nexus (and hence different versions of the national identity) was undergoing change during the 1950s. Arguing from the theoretical model introduced in Part 1, it is now possible to analyse the critical practices of the 1950s in terms of their engagement with received decolonising discourses, and also to consider the extent to which the postcolonial critic-as-subject might be able to move beyond or through the limitations of those received discourses towards critical practices which problematise and/or evade the derivative structures of traditional decolonising activity. While attempting to place a wide range of critical activities from the 1950s in terms of the model of decolonisation discussed at length in Chapter 1, this study will also proceed by way of a loose cross-fertilisation of a number of methodological categories, amongst which the most important are: (a) form; (b) analytical strategy; (c) institutional affiliation. Let me explain briefly what this will entail.

The various *forms* in which critical discourse appeared in the period will determine the format of the study. Chapter 5 looks at the periodical press of the time, concentrating mainly on the journal to analyse critical discourse along a continuum from scholarly/specialist-academic to popular/humanist-interventionist. (As this chapter introduces many of the issues peculiar to the 1950s, it will of necessity be somewhat longer than the following ones.) Chapter 6 examines the scope of literary criticism during the period, concentrating on important areas of non-standard 'critical' activity in the 1950s – pedagogy, censorship and anthologising – to see how a metadiscursive principle is invoked to aid or problematise processes of national subject formation in contexts beyond the usual notion of criticism as textual commentary. As was argued in Chapter 2, criticism is understood here not only as a set of reading strategies but also as 'a historically specific and institutionally grounded set of discursive practices which produces the literary as a space open to (different) kinds of intervention ...' (Bennett, 1990: 209) – that is, as any manoeuvre which looks to intervene in the relations between readers, texts and society. Finally, Chapter 7 will be an analysis of book-length studies, essays, theses, monographs, literary history, focusing especially on the reaction of Irish critics to the development of international Irish Studies and the implications of both (development and reaction) for Irish decolonisation. Such a survey does not aim or claim to be exhaustive. Rather, the intention is to give an impression of some of the significant ways in which critical discourse operated in Ireland at this time, the forms it took, the strategies it employed and the various functions it imagined for itself.

Another methodological strand employed attends the critical text's *analytical strategy*, and this enables a discussion of the various styles and techniques employed in the above-mentioned forms. The expansive definition of criticism employed here comprehends a wide range of activities, including all the usual processes associated with the practice: reading, responding, scholarship, explication, interpretation, exegesis, analysis, appreciation, discussion, evaluation and theory (Hawthorn, 1987: 22–31). The fate of certain critical analytical strategies over the decade is, I shall suggest, linked in significant ways to wider discourses of decolonisation and national identity.

Finally, critics will also be placed in terms of the *affiliations* they demonstrate towards one or other, or a combination of, prominent twentieth-century Irish institutions. A four-part model borrowed and adapted from Liam O'Dowd's analysis of Irish intellectual practices in the twentieth century will be employed, categorising critical subjects as: (i) state-affiliated; (ii) church-affiliated; (iii) literary; (iv) sectional (O'Dowd, 1985). This will provide a base from which to contextualise and compare the subjects who engage in critical discourse and the sorts of arguments and strategies they are likely to bring to bear on cultural material.

'After the departure of the imperialists, who are generally fairly recognisable,' writes Emer Nolan, 'the most important distinction to be drawn – and it is a much more difficult one – is between those who are complicit with neo-colonialism and those who are not, whether they be former natives or former settlers' (1994: 161). This book constitutes an attempt to draw that difficult distinction. The task is to examine the role played by literary criticism in the ongoing narrative of Irish decolonisation during the putatively 'postcolonial' 1950s, and to trace the provision of various institutional and social spaces wherein Irish culture has been enabled (or not) to express this narrative. Deploying the methodology described, some of the questions we might expect to address include: does a critical text display characteristics which enable us to identify it in terms of liberal and radical modes of decolonisation, or does it evince formal and/or conceptual awareness of the possibilities of problematising these categories through strategies of displacement, self-reflection or repetition? How does a critical text hold in tension a narrative line necessarily implicated in issues of 'Irishness' and 'non-Irishness' while looking to step outside such discursive limits? How did critics negotiate the relations between, on the one hand, institutional affiliation, form and analytical strategy, and, on the other, their engagement with the received narrative of decolonisation, whether the intention be to confirm or to refuse the terms of that narrative?

5 The Periodical

The value of the periodical publication has been recognised in Ireland since at least the eighteenth century, although it is only more recently that it has come to function as a major tactical weapon of Irish political and cultural debate (Kearney, 1988: 250–68). In the century after the first appearance of *The Nation* in 1842, a number of different types of journal appeared which could use the extraordinary success and effectiveness of that publication as a yardstick for their own aspirations. Some, such as Griffith's *The United Irishman*, Pearse's *An Claidheamh Soluis* or Connolly's *The Workers Republic* might desire *The Nation*'s ability to harness a mass readership and actively intervene in current debates. Some, such as Yeats's *Beltaine* and *Samhain* might covet the form's potential audience while wishing to maintain their esoteric cultural pursuits. Others, such as the various university and scholarly organs, might consider themselves above such populist activities, their work geared instead towards a small initiated readership who would appreciate empirical research and sustained analysis rather than the polemics and holistic declarations of their more accessible fellows. Some, finally, such as George Russell's *The Irish Statesman* and Seán O'Faoláin's *The Bell* might try to incorporate all the above into a discourse of cultural criticism, attempting to strike an eclectic balance between intervention, reflection and critique. Traces and developments of each of these approaches can be found in all subsequent Irish periodical literature.

These boundaries were extremely flexible, and certain individuals such as Russell could range between the various levels of discourse with ease (O'Dowd, 1985). Implicit in each, however, was an acknowledgement of the convenience and adaptability of the modern journal form as a means of engaging in up-to-the-minute exchange with colleagues and opponents. Whereas the major historical or literary treatise always risks being rendered *passé* by current events, the essay, the editorial, the work-in-progress and the review are present-oriented discourses, always provisional and placeable responses on the part of subjects locatable in time and space. They are *discursive* in the sense that they are recognisable interventions in ongoing debates, responses and interjections and rejoinders which imply other subjects and other points of view. Especially during the volatile years on either side of the revolution, Irish cultural and political opinion was constantly having to react to rapidly changing circumstances, and much in the way that the short story became the dominant literary form of the post-revolutionary period, so the periodical press became at this time a sort of halfway house between the newspaper and the book – neither journalism nor monograph but incorporating aspects of both – as a means for the Irish intellectual to intervene in the debate over national identity.

The period between 1948 and 1958, says Seamus Deane, 'saw the death of more journals than any other decade before or since' (1986: 229), and this is indicative of the constant changes impacting upon cultural debate at this time as certain points of view crystallised into publishing energy for a while, only to dissipate as that energy came to be seen as misplaced or irrelevant. The period also saw, however, the consolidation of certain forms of periodical literature affiliated to institutional discourses such as Celtic Studies, Theology and specialist scholarship in areas such as Classics and Philosophy. In this flurry of intellectual activity the principle of literary criticism was widely employed, as it had been since *The Nation* and before, as a means of analysing Irish identity. Literary criticism, that is, continued to be a major cultural and political preoccupation as the subject positions and narrative possibilities implied in different ways of reading texts implied different ways of being 'Irish'.

Richard Kearney suggests that the Irish intellectuals of the 1950s accepted and contributed to a formal split in journalistic discourse between, on the one hand, socio-political debate dealing with reality, and on the other, literary debate dealing with imaginative vision.[1] While this is certainly true to a point, it is far from the whole story. Such a neat division underplays the subtle negotiations and struggles over disciplinary borders and boundaries that were carried on in the pages of all the journals of the 1950s, as well as the power invested in certain discourses to pronounce on issues of national import. In the following pages I shall compare and contrast three major types of journalistic discourse of the period in terms of the model of decolonisation introduced in Chapter 1 and the loose methodology described in Chapter 4. Before that, however, as an introduction to the debate, one particular journal – *Kavanagh's Weekly* – will be examined in some detail to establish a frame of reference in which the other texts may be discussed. By then comparing a number of journals of the various types mentioned above which were available in the period between 1948 and 1958 (but especially over the thirteen-week life of *Kavanagh's Weekly*) it will be possible to assess the engagement of postcolonial periodical literary criticism with the ongoing issue of Irish decolonisation.[2]

The Moment of Kavanagh's Weekly

Kavanagh's Weekly appeared in thirteen issues between 12 April and 5 July 1952, in the middle of the period under investigation here, and at the heart of the economic and cultural malaise just described.[3] As one critic puts it, '*Kavanagh's Weekly* captures the *zeitgeist* of a particularly depressing period in Irish history. The country's economic boom was over and its spirits, like its finances, slumped. The atmosphere of apathy and pessimism was almost palpable' (Quinn, 1991: 279–80).

The journal was financed by Peter Kavanagh, who had recently returned from America with a sizeable amount of money. He found his brother living in poverty and squalor in his Pembroke Road flat with no outlet for his writings since the demise of John Ryan's little magazine *Envoy*, for which a monthly 'Diary' column had provided a regular if modest income. The journal was founded as a vehicle for the brothers' critique of contemporary Ireland, and with just a few contributions from other writers they produced over the next three months about 9,000 words a week. In a sustained onslaught against what they considered to be the faults of both nation and state, virtually no one and nothing, past or present, was spared. During their short run the Kavanaghs succeeded in offending large sections of the population as well as a number of powerful and sensitive individuals.

Patrick Kavanagh was forty-seven years of age in April 1952 and already an established poet and man-of-letters around Dublin. He had been 'found', as he said himself,[4] by AE in the 1930s, a primitive young country poet who had walked to Dublin from his birthplace (Mucker in Monaghan) on his first trip to the capital because he believed it to be in keeping with the peasant persona he liked to adapt in his poetry of that time (Quinn, 1991: 13). His reputation was made with the long anti-pastoral epic *The Great Hunger* published in 1942, and in the following decade he came to be regarded along with Austin Clarke as one of the major Irish poets since Yeats.

Like many other writers during the Emergency years and after, Kavanagh encountered the dearth of outlets for cultural discourse in Irish civil society, and his work suffered as a consequence. After

The Great Hunger there were only two other 'creative' publications before 1958, when the poems that would eventually comprise the volume *Come Dance with Kitty Stobling* (1960) were published as *Recent Poems*, once again by his brother Peter on his own printing press in New York. These publications were *A Soul for Sale* (1947), a collection of poems which contained an expurgated version of *The Great Hunger*, and a novel, *Tarry Flynn* (1948). This lack of a public for literature became one of the many indications in *Kavanagh's Weekly* of the failure of the new Ireland:

> The death which has overtaken Irish life in other fields has descended on the field of literature ... There is no use in concealing the fact that there is practically no literary public in this country and there has never been a literary tradition ... If a writer appeals to the few who count he may get all sorts of commissions. But it can be taken as a fact that no sincere writer can make a living by his creative writing. (3:7)

Kavanagh immediately establishes a link between 'life' and 'literature', but then goes on to criticise the way this link has developed in post-revolutionary Ireland. The 'thin society' could not morally sustain the narrative complexities of Irish literature, or financially support the individuals who might wish to write that literature. Instead, the postcolonial intellectual was absorbed by various state or semi-state institutions, such as Radio Eireann, the Abbey Theatre, the Universities, the Dublin Institute for Advanced Studies and the Cultural Relations Committee. The alternative was to scrape a precarious existence on the hackwork available at home or, if one was as famous as Seán O'Faoláin or Frank O'Connor, on occasional commissions to the journals and publishing houses of London and New York. For a sincere creative writer like Kavanagh, recourse to state-institutional outlets for his work was demeaning, at least while *Kavanagh's Weekly* was still a viable concern; and although he accepted a temporary lectureship at University College, Dublin a few years later, the point about the difficulty of making a living through imaginative writing was well taken by the Irish poets, playwrights and novelists of the 1940s and 1950s.

Although *A Soul for Sale* and *Tarry Flynn* were reasonably successful, by far the greater energy at this time was expended by Kavanagh on journalism, reviewing and, when *Envoy* came along, on that mixture of subjective pronouncement and literary assassination that was to become so familiar to readers of *Kavanagh's Weekly*. For example:

> The play at the Olympia, *The Land is Bright*, is about as malignant a tumour as has ever been grafted on to Irish village life. It is almost more

> indecent in its idiocy than a L.A.G. Strong job ... The play is definitely a tragedy – for the audience. It is not to be wondered that the author preferred to remain anonymous, a fact which should be considered as it shows that he is not so devoid of sensibility as might at first be thought ... I often wonder if Ireland isn't utterly inferior, an exhibitionistic society whose natural gods are actors. (2:8)

In such a highly litigious society as post-independence Ireland Kavanagh's 'honest' critical style was dangerous and exciting, and had already earned him a reputation as an unsafe journalist.[5] No one was protected from the relentless search for 'truth', and neither delicacy of expression nor subtlety of opinion was required by one who was known 'to treat all literary allegiances, friendships or truces as temporary' (Quinn, 1991: 459). His reasons for disliking *The Land is Bright* probably had to do with authenticity of representation and his general dislike of the Dublin theatre scene, but no attempt is made to analyse the play in any depth or to defend his evaluation of it. Instead, the pejorative cultural judgement is immediately linked to a comment on Irish society, and cultural and political failure are thus mutually implicated.

Indeed, throughout *Kavanagh's Weekly* this implicit relationship between cultural and more general social and political concerns is stressed. Like his literary foster-father AE, during this stage of his career Kavanagh took all Irish experience as his province, and throughout the thirteen issues of the journal he ranged freely over a wide variety of discourses, including economics, politics, popular culture, sociology, philosophy and psychology. The full title of the publication is *Kavanagh's Weekly: A Journal of Literature and Politics*, but although the cultural and political aspects are formally separated into articles and editorials, the search for truth and falsehood in all their forms remains constant throughout. The poor state of Irish literature is never far away in an editorial about the state of the nation, while there is always a link, sometimes stated, sometimes implicit, between politico-economic performance and cultural vitality. And this is as true of the old Gaelic society as it is of the modern Irish one:

> The Irish tradition regarding the poet was not a good one; it constantly and persistently encouraged the poet into undignified ways. The real trouble is that perhaps there was no civilised tradition ... The Gaelic poets hardly deserve the name of poets for they lacked the one quality of a poet – leadership. A poet draws the people's attention to the obvious that they otherwise do not see: he gives them courage, the courage to be themselves – the only kind of courage that is worth having. (10:7)

The contradictions between community and personality that were to dominate Kavanagh's thought and work for the rest of his life are already apparent here, and his first tentative attempts to achieve a resolution of these categories will be analysed shortly. Also worth noting, however, are: firstly, his harnessing of poetic and social ('civilised') traditions, often collapsed into the abstract quality of 'Life' (as he wrote: 'Art is never art. What is called art is merely life' (6:7)); and secondly, the implicit truth that the writer-critic is the individual in society best qualified to diagnose this relationship. Like other post-Revival writers such as O'Faoláin, O'Connor and Clarke, Kavanagh coveted the kind of affective aura possessed by a Yeats or an AE. However, in an age of increasing specialisation and professionalism, the day of the 'traditional intellectual' (Gramsci, 1971) was nearing its end, and the kind of amateur, interventionary power desired by many of the intellectuals of the 1950s was no longer available (O'Dowd, 1985, 1988).

Kavanagh's critical enterprise was unusual, then, in that unlike most critics in the twentieth century he had little experience of formal academic discourse, and it was not until the mid-1950s that his critical work received any institutional patronage.[6] His affiliations were always with literature first and then with that section of the population he believed had been criminally misrepresented by the writers and idealogues of both revival and revolution – the small farmers, agricultural labourers and village dwellers of rural Ireland. *Tarry Flynn* he claimed to be 'not only the best but the only authentic account of life as it was lived in Ireland this century' (Kavanagh, 1987: 394). These affiliations – cosmopolitan and sophisticated on the one hand, local and simple on the other – gave a curious contradictory tone to many of his pronouncements on Irish life and literature; indeed, it is the ambiguity of his stance that makes Kavanagh such an interesting figure in terms of his critical engagement with decolonisation. The targets of his attacks were invariably those official church- and state-sponsored institutions and those individuals – especially de Valera and Austin Clarke – whom he believed guilty of perpetuating disabling myths about Ireland's present and recent past. Editorial control allowed Kavanagh the freedom to vent the anger he felt at the role of poor neglected poet to which modern Ireland had condemned him, and to attack those institutions and individuals he believed to be responsible for this fate; at the same time it was also a forum where he could begin to formulate an aesthetic with which to negotiate the particular set of circumstances in which he found himself. Like many other post-independence intellectuals, Kavanagh experienced the tension between alienation from and commitment to the new Ireland, and this split between destructive political iconoclasm and constructive cultural utopianism is clearly visible in the pages of *Kavanagh's Weekly*.

The destructive element is founded on Kavanagh's belief in his own ability to recognise the reasons for Ireland's failure and his willingness to inform as many people as often and as honestly as possible what those reasons were. Interestingly, Kavanagh formulated this capacity as 'the critical mind':

> Readers unused to the critical mind may think us destructive but this is because they are accustomed only to the flabby unthinking world of the popular newspaper.
>
> Our hope is to create in a few thousand people the power to think critically before it is too late. In life there needs to be a constant battle to recover losses. Even to hold your place you have to fight. Hence what looks like destructiveness is merely the critical mind ... (3:1)

Again, those members of society more sensitive to 'life' were more likely to be in possession of 'the critical mind' than others, hence the embattled, missionary tone of much of the writing in *Kavanagh's Weekly*. But what is also interesting is the fact that 'the critical mind' can be brought to bear on both cultural and political phenomena; it is a transposable faculty that can allow the initiated to see beneath the false mask worn by modern Ireland. Literary criticism, then, must be consistent with cultural, political and social criticism, as in each case it is the same cross-fertilisation of 'the critical mind' with 'life'. And as all the above extracts indicate, modern Ireland was characterised for Kavanagh by a shameful lack of either of these qualities.

Kavanagh's negative criticism often threatened to dominate his personality and work, leaving him, as Antoinette Quinn writes, '[a] man flailing between two faded worlds, the country he had left and the literary Dublin he never found' (1991: 283). Kavanagh's solution to the aesthetic and personal dilemma caused by modern Ireland was based, famously, on his discovery of the concept of parochialism and its use as a metaphor for gaining access to poetic and social truth. 'The job', he wrote, 'is to find some substitute for the national loyalty, some system to take the place of the enslaving State' (10:1). Writers and critics need not look abroad, or even at the larger community, for their subjects, he claimed, but only at the things which directly impinge on their own experiences:

> Parochialism and provincialism are direct opposites. The provincial has no mind of his own; he does not trust what his eyes see until he has heard what the metropolis – towards which his eyes are turned – has to say on any subject. This runs through all activities.
>
> The parochial mentality on the other hand is never in any doubt about the social and artistic validity of his parish. All great civilisations are based on parochialism – Greek, Israelite, English ...
>
> In Ireland we are provincial not parochial, for it requires a great deal of courage to be parochial. When we do attempt having the courage of our parish we are inclined to go false and to play up to the larger parish on the other side of the Irish sea. In recent times we have had two great Irish parishioners – James Joyce and George Moore. (7:2)

Here, Joyce rather than Yeats is the touchstone of authentic Irish art because the work of the former is thoroughly informed with the local, with 'life' as it could still be experienced on the familiar streets of Dublin. 'Yeats, protected to some extent by the Nationalistic movement, wrote out of a somewhat protected world and so his work does not touch life deeply. Joyce on the other hand got all that Ireland had to give and his reaction to it made him great' (6:7). More than any writer or critic since the Revival, Kavanagh managed to breathe life into what in the 1950s was threatening to become an exhausted opposition between nationalism and cosmopolitanism, on the way creating 'something new, authentic and liberating' (Heaney, 1980: 116). For Kavanagh, parochialism did not connote insularity or essentialism as did the version of nationalist ideology dominant since independence; instead it signified a capacity to hold the local and the universal in fructifying tension. For example, *Kavanagh's Weekly*'s blatantly anglophilic stance was not only a goad to 'buckleppin'' nationalists (for which term see below), but also a genuine wish to bring Empire and parish into mutually enlightening confrontation. The seemingly contradictory desire of the Irish intellectual for what Seamus Deane has called 'the risks of modern individuality and the consolations of traditional community' (1986: 213) was not, for Kavanagh, a limit for either himself or Ireland, but a potentially emancipatory displacement of categories which had traditionally shored up dominant nationalist discourse. This acknowledgement of the universal in the local, the specific in the general, allowed Kavanagh to maintain a subversively ironic attitude towards life and literature as he experienced them in postcolonial Ireland. As such, the concept of parochialism developed by him represents one of the first major attempts by an Irish intellectual since Joyce to introduce a qualifying perspective into the narrative of decolonisation.

Parochialism, as well as indicating a lifestyle, also demanded a literary style that would be capable of articulating the local community and the role of the poet, and it was at this time that Kavanagh began to develop the aesthetic that would later emerge as the relaxed, conversational tone of the *Kitty Stobling* verses.[7] In attempting to maintain a consistent persona throughout his political, critical and poetic discourse, Kavanagh developed a highly present-oriented style founded on personality and immediacy. Whether the subject be the state of the economy, gambling on horses or an anthology of Irish poetry, the commenting voice should remain the same, always searching for and exposing the 'truth' behind the various forms of lies on which modern Irish life was built. That voice could float in and out of 'poetic' and 'prosaic' registers without warning, as in this 'criticism' of schoolbook poetry:

> For me, when I read Eugene Aram, I am back in my native place, aged about sixteen with all my dreams sealed in the bud ... There I am walking down a lane peeping through the privet hedge into the field of turnips. The mood and atmosphere of the time comes alive in my mind. The comfortable worry of the summer fields is upon me. All the bits and pieces that furnish Imagination's house come up by magic ... I am walking through a field called Lurgankeel away down towards a shaded corner; it is an October evening and all around me is the protecting fog of family life. How shall I live when the fog is blown away and I am left alone, naked? (5:8)

The sentimental, interrogative tone of the above is at odds with the realistic, assertive style of most of the writing in *Kavanagh's Weekly*. It thus unsettles the reader's expectations about the appropriateness of certain kinds of writing to certain subjects, defamiliarising social and cultural norms at the textual level. Equally, the temporal location of the critical voice in *Kavanagh's Weekly* is always problematic, with the present circumstances constantly threatening to impinge on the discourse, thereby giving all the pronouncements a provisional, spontaneous air. For example, an article on the contribution of George Moore to modern Irish literature digresses into something out of a Myles na gCopaleen 'Cruiskeen Lawn' column:

> I started off to write about George Moore. Perhaps I should forget about him and concentrate on the train of thought that his name provokes ...
> – A casual passer-by: What about George Moore?
> – Forget about George Moore, there's another week. (7:7)

Or in speculating on the influence of Boswell on Joyce, Kavanagh quotes similar diary entries from *Boswell in Holland* and *A Portrait of the Artist as a Young Man*, then adds the following:

> March 21. Thought this in bed last night but was too lazy to add to it. Boswell was twenty-two, the same age as Stephen, at the time of the writing. Feel odd similarity. Had Joyce seen Boswell's diary? No. Couldn't have. Considering above wonder if it adds anything to our knowledge of anything or anybody. Probably not. (5:8)

The ironic, *ad hoc* and playful nature of these last two extracts is indicative of much of the literary criticism of *Kavanagh's Weekly*, and it is clear that the critical style to which the philosophy of parochialism gave rise would be of a highly idiosyncratic kind, having little to do with scholarship, explication or analysis, but much to do with subjective response and evaluative assertion. By unsettling the borders between his cultural and literary discourse, Kavanagh attempted to avoid the all too real danger for the postcolonial Irish writer of letting a political conscience smother an artistic impulse (Harmon, 1984: 56ff.).

The first editorial, entitled 'Victory of Mediocrity', is a good example of the general style and approach of *Kavanagh's Weekly*. It is a belligerent revisionist interpretation of the achievements of the revolution, going on to attack openly the individuals and institutions responsible for perpetuating the lies on which the thin society of postcolonial Ireland eked out its miserable social, economic and cultural life:

> Thirty odd years ago the southern section of this country won what was called freedom. Yet from that Independence Day there has been a decline in vitality throughout the country. It is possible that political liberty is a superficial thing and that it always produces the apotheosis of the mediocrity. For thirty years thinking has been more and more looked upon as wickedness – in a quiet way of course.
>
> All the mouthpieces of public opinion are controlled by men whose only qualification is their inability to think.
>
> Being stupid and illiterate is the mark of respectability and responsibility. (1:1)

Already the reader is introduced to what were to become some of the keywords of *Kavanagh's Weekly*: vitality, mediocrity, responsibility. The style is straightforward and accessible, the tone assertive and angry, the persona oscillating between that of the informed social/political commentator and the plain man of the people, telling it as he sees it. Kavanagh never patronises, panders or condescends to his readers; farmer and academic, peasant and politician, all are equally addressed and implicated in contemporary Ireland's dilemma. What it also demonstrates at this early stage is a willingness to question all the images and narratives which had emerged in Irish cultural discourse during the process of decolonisation. Later in the same editorial Kavanagh refers to the liberty that was won by the previous generation as 'the liberty of the graveyard' (1:1), and throughout the thirteen issues of *Kavanagh's Weekly* a highly ironic attitude towards all the national shibboleths is maintained.

Each week brought a literary critical article in which Kavanagh raged against what he considered to be the depressing state of contemporary Irish letters. His main targets were the 'buckleppers' – enemies of the creative imagination who made a living from selling false images of Irish life. The grandfather of all 'buckleppin'' literature was Synge, but it was his modern-day disciples who week after week bore the brunt of Kavanagh's ire – writers such as F.R. Higgins, W.R. Rodgers, Walter Macken, Mary Lavin and Austin Clarke. In a piece entitled 'Paris in Aran', after accusing Synge of transposing cosmopolitan attitudes into the rural environment of the West of Ireland, he continues:

> What is the dominant note in Synge? I would say bitterly non-Irish. It all came from the basic insincerity upon which he built ... Synge never asked himself the fundamental question: where do I stand in relation to these people? Whether or not Synge portrayed the people of the West truly is not of much importance; as I say, it is the lie in his own heart that matters ... His peasants are picturesque conventions; the language he invented for them did a disservice to letters in this country by drawing our attention away from the common speech whose delightfulness comes from its very ordinariness. (9:7)

In an argument that goes some way along the road mapped by Corkery twenty years earlier, Kavanagh sees Synge's major flaws as his distrust of ordinary Irish life and language, and his inability to gain access to authentic Irish experience. Moreover, the insincerity of representation that resulted in Synge's work was being compounded in modern Ireland by 'buckleppers' such as the Irish branch of the PEN, whose anthology of Irish poetry, entitled *Concord of Harps* (from which he was excluded), Kavanagh reviewed as 'a piece of satire that couldn't be equalled by a man trying to satirise', and 'the funiest [sic] book I have read for years' (4:7). Where Kavanagh parted company with Corkery (and with most other social commentators of the previous and present generations) was in his understanding of 'authentic Irish experience', which, as we have already seen, he characterised as an unstable compound of life, personality and parish.

Against the Synge-inspired buckleppers Kavanagh set what in one article he called 'saints of the imagination' (2:7): Joyce, and to a lesser extent Yeats. 'One phrase of Joyce', he claimed, 'is worth all Synge as far as giving us the cadence of Irish speech' (9:7), while he considered *Ulysses* one of the two greatest works (along with *Moby Dick*) in English prose fiction (12:7). The 'Joyce' of his pantheon, however, would have been unrecognisable either to those who relied on ethnicity as a critical criterion or to the scholars, exegesists and theorists who at this time were embarking on the international institutionalisation of Joyce's work. *Ulysses* was not a career for Kavanagh but a striking example of parochialism in action, a book impregnated with Dublin but at the same time 'only incidentally about Dublin and fundamentally the history of a soul' (12:7). He valued (and desired), in other words, Joyce's local interventionary potential and his ability to bring parish and personality to life in his art. Kavanagh's admiration of Joyce was not that of the critic for the subject, but the writer-critic's acknowledgement of the power of another writer.[8]

Seldom does Kavanagh sustain a line of argument throughout a whole article. Instead he digresses rapidly from pseudo-analysis to pseudo-theory to evaluation and back again in the course of a few hundred words. The effect is more like a pub monologue that a reasoned textual response – rambling, anecdotal, impulsive and occasional. It is by making his critical discourse into a spontaneous, situated, occasional event rather than a timeless, voiceless text that Kavanagh attempts to insinuate a role and a power for himself in a society which, turning increasingly towards professionalism and institutionalism, was rapidly doing away with the very notion of the holistic writer-critic towards which he aspired.[9]

Neither as business enterprise nor as cultural intervention was *Kavanagh's Weekly* a success.[10] Writing large portions of each issue by himself and maintaining the same tone and style throughout each page of each copy, while strategically subversive and progressive, was tactically counter-productive for Kavanagh in that the journal was very one-dimensional, its note sounding in a 'shrill monotone' (Quinn, 1991: 284) rather than the melodic counterpoint with which the periodical form works best.[11] However, with the development of the concept of parochialism, the attempt to expose the partial vision on which dominant nationalist ideology was constructed, and the resistance offered to the hegemony of specialisation in the realm of literature, *Kavanagh's Weekly* represents an important moment in the narrative of modern Irish decolonisation as mediated by the discourse of literary criticism.

Variations on The Bell

What other kinds of periodical literature were available between April and July 1952 when *Kavanagh's Weekly* was laying waste the modern Irish cultural and political landscape? One type of publication was the magazine predominantly concerned with new creative literature but which also incorporated a certain amount of editorial comment, book reviewing and criticism in which the aesthetic ideology behind the 'primary' imaginative literature could be framed and defended. This kind of discourse was undertaken by figures not unlike Kavanagh, although the writer-critics of *Poetry Ireland*, *Irish Writing*, *Envoy* and *Rann* tended to encourage a self-conscious avant-garde aura that would have dismayed the more established poet. Kavanagh, as seen above, relied for effect not on formal or intellectual innovation but on consistency of voice. The major difference between *Kavanagh's Weekly* and these journals was that the sectional consciousness of the former – the sense of a social and political context for its literary affiliations – was generally less developed, if not completely absent, in the latter. These journals were generally aimed not at the wider public but at a limited group of intellectual readers. This may initially appear to be a strange observation if one accepts that the main influence on this kind of publication, if not actually a particularly good example of the genre, was *The Bell*.

Like *Kavanagh's Weekly*, *The Bell* 'stands as the record of an alternative culture' (Foster, 1989: 548) in postcolonial Ireland. It began publication in 1940 under the editorial control of Seán O'Faoláin. Although in its first run (which ended in 1948) it had been subtitled 'A Magazine of Creative Fiction', *The Bell* was never solely a literary undertaking but, like AE's *The Irish Statesman* before it, interested in reflecting and intervening in all aspects of contemporary Irish life.[12] Along with its dedication to 'Life before any abstraction' (O'Faoláin, 1:1:1) and its consolidation of the role of writer as social critic, *The Bell* was thus also an influence on *Kavanagh's Weekly*. However, the two dates already mentioned in connection with this periodical – 1940 and 1948 – are significant in that, although not separated by a great number of years from 1952, they represent the experience of a different generation from

Kavanagh's Weekly. O'Faoláin, after all, had participated in the military struggle, and in 1940 was still close enough to the actual events of the revolution and its aftermath for those events to bear significantly on his intellectual profile. April 1948, when the first run of *The Bell* ceased, predated the announcement of Ireland's withdrawal from the British Commonwealth and the decision to become a Republic; the early numbers were thus engaged with critical debates structured in terms of received (liberal and radical) discourses of decolonisation. When it recommenced publication in November 1950 under the editorship of Peadar O'Donnell, therefore, *The Bell* had a legacy of working, albeit progressively, within certain discursive parameters dictated by Ireland's recent revolutionary history, and O'Donnell (another former activist from the revolutionary period) found it impossible to imagine an alternative editorial agenda. The journal disappeared for good in 1954.

Although they overlap in time, in fact, *The Bell* and *Kavanagh's Weekly* differ in many significant ways – editorial style, selection of contributors, textual format, price, the attraction of commercial advertisement, length and so on.[13] One important difference was that, unlike *Kavanagh's Weekly*, much of the content of *The Bell* was imaginative literature. During the period between April and July 1952, the latter included poems and short stories by important contemporary figures such as Padraic Fallon, John Hewitt, Liam O'Flaherty, Seán O'Faoláin and Kate O'Brien. Indeed, throughout its run it attracted contributors unlikely to have been found, by desire or by editorial invitation, in the pages of *Kavanagh's Weekly* (Holzapfel, 1970).

The major difference between the two publications, however, emerged in *The Bell*'s critical discourse and the different function it imagined for the critic and the artist in relation to the nation. Criticism was a way of simultaneously advertising Ireland's acceptance of the modern world and its right to participate in the universal discourse of letters. By engaging in what it imagined to be normal international critical business-as-usual (that is, as carried on in other modern democratic cultures such as Great Britain and the United States), *The Bell* could affect the discourse of (thereby helping to create) a traditional relationship between artist, critic, civil society and state in modern Ireland. In this sense, *The Bell*'s critical agenda was still being set by the events of 1916–22 as it sought to diagnose and defend the revolution under the gaze of, and in terms recognisable to, non-Irish subjects. The critical practices found in *The Bell* reveal that its editorial policy was still operating in terms of that double bind of sameness and difference which characterised the dominant modes of decolonisation.

For example, *The Bell* was much more conventional than *Kavanagh's Weekly* in its reviewing practice. Each month a number of books from different disciplines would be discussed briefly with regard to content and merit by a number of suitably qualified individuals, a format that may be traced back to the literary and cultural journals of the late eighteenth century. The period between April and July, 1952, when *Kavanagh's Weekly* was running, included reviews by John Hewitt on *Rainer Maria Rilke: His Life and Work* by F.W. Van Heeri Khuizen (18:1:55–6); Geoffrey Taylor on *The Background of Modern Poetry* by J. Isaacs (18:2:119–20); and Austin Clarke on *The Letters of Ezra Pound* edited by D.D. Paige (18:3:189–90). These were all distinguished contemporary Irish intellectuals writing competently, often perceptively, on a variety of subjects related to literature – not 'Irish' literature but the universal phenomenon of writing as it related to the wider international community. Evaluation was not abandoned, but was generally disguised as a form of literary appreciation. The criticism was conducted in an urbane, humanist style which at its best could convey intelligent insight and comment, but frequently strayed into petty *bellettrism*.

This is true also of the longer critical pieces which were often included in *The Bell* – essays, diaries, extended reviews and transcribed lectures. Thus, while both Kavanagh and a writer-critic such as Padraic Fallon might share a dislike of the contemporary Dublin theatre scene, the latter was likely to cast this critique in terms as reprehensible to the former as the worst dramatic practice imaginable:

> I am not drawn very much to our modern stage because it has, in Dublin at any rate, frozen itself into a three-walled convention where there can be only a theatrical poetry of situation and no metrical poetry, no verbal enlightenment such as a chorus can give. A poet needs to break down those three walls somehow, and let another world through so that, indeed, some of the characters bring a comet-tail of something larger than the human along with them. (18:2:109)

Some parts of Christopher Fry apart, modern poetic theatre was anathema to Kavanagh. Its Irish practitioners and apologists (such as Austin Clarke and Fallon himself) called forth some of his most splenetic outbursts (*Kavanagh's Weekly*, 5:6, and *passim*). The smug, informed tone of the above, the affected 'indeed' which punctuates the last sentence, and the quasi-Romantic sentiment of the whole would all have been grist to Kavanagh's iconoclastic mill, apparently proving his thesis that 'life' had been effectively removed from not only the political but also the cultural agenda of contemporary Ireland.

Especially in its second run, *The Bell* faced the problem of finding a market for itself. As the extract from Fallon shows, the criticism implied if not a specialist audience, at least a certain kind of informed readership, mostly subjects who engaged in, or were interested in, creative writing. However, this was a fairly narrow market, as there were other publications which dealt solely or predominantly with creative writing, while its lack of affiliation to any educational or professional institution meant that *The Bell* possessed no specialist languages or discourses to attract a regular readership. This dilemma emerges, for example, in a piece entitled 'The New Criticism' in the June 1952 issue in which O'Faoláin demonstrates the power of relating international developments to local experience that made him such an important figure in post-independence intellectual circles:

> Wherever else art is today, it is in the market-place, the pub, the queue, the bus, the office, parliament and the press, in the kitchens and the hospital, as happy painting frying pans as writing of potato stalks. Art is no longer something separate; it is what it always was in the great ages ... a vital part of common life. It is only the academicians and the academic critics who are still treating it as a luxury. Why does not the New Criticism see that it is playing into the hands of these gentlemen by so rigidly separating the poem and the poet – the created thing from the man and all his life and all of our lives of which it is not just a reflection and recreation but an actual part? (18:3:141–2)

There are two interesting points to note about this extract and the article from which it is taken. The first is that here again, as in much postcolonial intellectual discourse, the theme is literary criticism itself and the insights it affords on the state of the nation. The Irish subject's reaction to 'New Criticism' can somehow tell us more about that subject than empirical research or even Art itself. The critic, located somewhere between the artist and the scientist, is the individual best placed to understand the ideas on which society relies to succeed; and for O'Faoláin, Irish society should not be any different from British or American or French society. The power of O'Faoláin's discourse lies in its affectation of the sort of holistic insight possessed by pre-revolutionary intellectuals such as Yeats or AE. In the postcolonial era, however, this Arnoldian strategy (of which O'Faoláin himself is one of the foremost practitioners[14]) is in itself, regardless of immediate local insights, a tacit acknowledgement of metropolitan cultural leadership, as 'Irish' experience looks to comprehend itself in the 'universalist' terms made available by 'non-Irish' sources. With the 'thin society' affording so little material for creative work, critical discourse turned inwards and began to contemplate its own conditions of existence; in that very movement, however, literary criticism engaged with the oppositional structures of decolonising thought which 'postcolonial' Ireland found so difficult to shake off.

Returning to the immediate context: who are the intended readers of this piece of literary criticism in the Ireland of June 1952? Here we find O'Faoláin rehearsing many of the themes which were emerging at precisely the same time in the pages of *Kavanagh's Weekly* – the importance of cultural discourse to the national health, the relationship between art and life, the necessity to resist the hegemony of the cultural institutions. O'Faoláin's style, however, is much more controlled, his tone less aggressive, the argument more coherent and polished. This is because his own point of view is a lot closer to the 'academicians' whom he is criticising than is Kavanagh's, and the gap between his literary affiliation and any sectional affiliation he might presume is as a consequence harder to bridge (Harmon, 1984: 190–7). In an argument reminiscent of Yeats fifty years earlier, O'Faoláin encourages Irish artists to pay attention to the New Critic's emphasis on 'technique' if they wish to convey adequately the 'mystery' of Irish life in artistic form; and as in much of the early Yeats, the tension of trying to balance cosmopolitan/metropolitan 'technique' and native 'mystery' is perceptible beneath the calm facade of the text's reasoned, reasonable position.[15] *The Bell* operated on the interface between holism and professionalism, indulging in general cultural pronouncement and quasi-social-scientific research, and in this way it *created* rather than *reflected* – presaged rather than reported – the conditions for the kind of 'Irish' culture it valued.[16] Talk of specialist intellectual discourses such as 'New Criticism' sits uneasily with references to pubs, queues and kitchens, and there is no disguising a certain patronising note in this piece as the learned discourse confronts, flatters and finally comes to dominate the 'common life'.

The problems of audience and ideology were also ones which confronted the journals and little magazines influenced by the format and relative success of *The Bell*, if some way removed from it in editorial intention. Two such publications which brought opposing solutions to these problems were *Envoy: An Irish Review of Literature and Art*, edited by John Ryan and Valentin Iremonger in Dublin, and *Rann: An Ulster Quarterly of Poetry and Comment*, edited by Barbara Hunter and Roy McFadden in Belfast. While both journals adopted *The Bell*'s format of combining imaginative, critical and editorial discourses, both were far removed from it (and from each other) when it came to conceiving of the function of literature and criticism in modern Ireland.

Envoy constantly restated throughout its short run a philosophy which made it a much more attractive proposition to those younger writers and critics alienated from the specifically nationalist problematic which seemed to dominate the pages of *The Bell*. One such editorial puts it this way:

> The younger poets ... take their nationality rather more for granted. They seem to be less interested in the technical craftwork by means of which one apparently becomes Irish – the over-use of assonance, Larminie, Raftery, a strained and imprecise Imagery – and rather more concerned with the craftsmanship involved in trying to write good poetry; their attitude seems to correspond more with Paul Gerrard's – that if the poet happens to be Irish, the result, as like as not, is probably Irish poetry. They would probably claim that being Irish is no more an attitude of mind than the wearing of embroidered coats. (3:9:6)

Although *Envoy* ceased publication in July 1951, this concept of incidental nationality was an important rejoinder to *The Bell*'s progressive nationalism, while its outward-looking stance was a strong influence on *Kavanagh's Weekly*. The editorial line held that literary affiliation took precedence over any kind of sectional or national affiliation, and 'literature' needed no qualifying adjective to make it valuable or interesting. Published in the bohemian quarter of Dublin and commissioning the more innovative and daring writers, *Envoy* possessed all the attractions of the new kid on the literary block, and to a younger writer like Anthony Cronin (who was involved with both it and *The Bell*), *Envoy* was both more relaxed and more exciting than its strait-laced older rival ensconced across the river among the banks and building societies of O'Connell Street (Cronin, 1976). In both the concept and execution of a special Joyce number in April 1951, for example (including poetry, letters, scholarly exposition and biographical reminiscences as well as a debunking editorial [5:17:6–11] from 'Brian Nolan'), *Envoy* seemed to be willing to confront head-on issues which had been skirted or alluded to only briefly in *The Bell*.

It was during his time with *Envoy* that Kavanagh, who contributed a 'Diary' column to all twenty numbers of the magazine, honed his contempt for the 'buckleppers' of modern Irish literature, although he would eventually reject *Envoy*'s unwillingness or inability to ground literary affiliation in some section of the national community. The question of finding some idea of community into which literary phenomena could intervene was also one which greatly exercised writers from the six-county statelet of Northern Ireland during this time. One favoured solution, and one which arises time and again in the pages of *Rann*, was that of regionalism.

The concept of regionalism was a recent importation from the United States where in the years after the civil war, those individuals and institutions lacking any affiliation with the corporate state reacted to their marginalisation from the centres of intellectual discourse on the east and west coasts by stressing the uniqueness of experience and tradition in the southern and mid-western states (Hoffman, Allen and Ulrich, 1947: 128–43). Transposed into the context of Northern Ireland in the 1950s, regionalism was well equipped to supply the writer with a concept of community that escaped identification with the larger corporate communities to the south and east. It was a unique, albeit fragile, solution to a fluid political situation, and one which, as long as it was never tested against a practical political crisis, provided writers such as W.R. Rodgers, John Hewitt, George Buchanan and McFadden himself with a source and an imagined audience for their work. In the Summer 1952 issue Buchanan asserted:

> I am, I hope, in the direct tradition of these Irishmen to whom I have paid both public and private homage. Think, also, of Swift. The Anglo-Irish tradition, in a word, differs from the English.
>
> It is the sense of world-life coming through a region that is, in my view, the principal interest, not a region hardening its shell against the universal life. A region, like a person, becomes valuable in so far as it transcends itself. (16:22–23)

Here, Buchanan specifically identifies with an 'Anglo-Irish tradition' which, since it can no longer find corporate or institutional expression in the Republic, is available for appropriation by those writers who (like the tradition itself) are spiritually and politically homeless. Yeats, Joyce, Shaw and Swift were clearly not English writers, but given the unbridgeable gap between the implications of their work and the current condition of that state which claimed the national nomenclature, neither could they be considered 'Irish'; therefore, they could be imaginatively transposed into a 'tradition' for the region, *Irish* writers who wrote in *English* but denied affiliation with either of the official expressions of those national signifiers.

In this way *Rann* invented an affiliation to a section which had no existence in fact. The journal's 'regionalism', as defined in the second paragraph of the previous quotation, while appearing to bear many similarities to Kavanagh's 'parochialism', in fact bore all the hallmarks of what he comprehended as a 'provincial' publication. *Rann*'s editorial tone was always defensive, the critical style it fostered always evincing that combination of resentment and deference which for Kavanagh characterised the provincial mentality. Roy McFadden's vigorous attempt to appropriate AE for the Ulster tradition in the issue of Spring 1952 is a good example of such a mentality in action; although Russell may have 'left the North for Dublin at the age of ten, he never forgot that he was an Ulsterman ... Yes, he was an Ulsterman: in accent, and also in that flinty streak of commonsense that made the mystic also the man of affairs, the organizer, the lecturer' (15:7). Both Irish nationalism and English literature are fended off here as McFadden attempts to explain AE's personality and work in terms of certain regional characteristics.

In other words, whereas *Envoy* attempted to offset the debilitating effects of a calcified nationalism by denying the culture/nation nexus and retreating into pure literary affiliation, *Rann* attempted to reformulate that link as 'culture/region', a tactic which left that journal in the curious position of having to invent *and* defend a living tradition simultaneously. Both publications, therefore, are enmeshed in debates characteristic of earlier decolonising discourses, their orientations and affiliations formed on premises of similarity to (*Rann*) and difference from (*Envoy*) pre-established national categories.

The same is largely true of *Irish Writing* and *Poetry Ireland*, periodicals endeavouring to intervene (as their titles suggest) in debates about the nature of contemporary Irish literary experience. The former began publication in Cork in 1946 under the editorship of David Marcus and Terence Smith; it continued until 1957 with Seán White taking over for the last three years. The second, also published in Cork and edited by David Marcus with John Jordan, appeared as a supplement to the former in 1948 and continued as a sort of sister publication until 1955. The format was similar in both cases, the difference being that one focused on shorter and extracted fiction and one on poetry, with the imaginative material in both supported by a small amount of critical material in the form of editorials, short critical essays and reviews. Despite the endemic financial worries, both publications were relatively successful and during the period under consideration here – Spring and Summer 1952 – they included work by writers such as Norah Hoult, Eric

Cross, Roy McFadden, Bryan MacMahon, Temple Lane, Thomas Kinsella and Ewart Milne.

Both journals shared *Envoy*'s dissident stance *vis-à-vis* nationalist history, agreeing that 'the general character of Irish literature, or more accurately, contemporary Irish writers, are only too happy to take their historical background for granted simply because in their youth they had it rammed down their throat'.[17] With the country undergoing a post-Emergency revival of 'buckleppin'' and the forces of reaction in state and church well positioned to contain any innovatory thinking, such a stance was a radical, even dangerous, opinion in the 1950s. Even more so than *Envoy*, however, the affiliations of these journals were narrowly literary (O'Dowd, 1988; Kearney, 1988), and despite the 'Irish' and 'Ireland' of their titles, both were geared towards an artistic model considered in universalist rather than local terms. Questions about the socio-political context of Irish literature, or even 'literature', if not strictly off-limits, were at least in slightly bad taste. In style and tone the literary criticism of both *Irish Writing* and *Poetry Ireland* tended towards *bellettrism* and a low-key neutrality. Thus, in the article on the modern Irish novel already cited, David Marcus writes of the need 'for another Joyce, but this time a Joyce unblistered by the Irish writer's "natural egotism"; a plain man's Joyce' (18:49). Here, Joyce's radical intervention into Irish cultural discourse is evacuated in terms redolent of essentialism and reaction. The Joyce constructed by Marcus is apprehended solely in literary terms and tamed for purely literary purposes – that is, the self-sufficient trajectory of the Irish novel. Or again, in a review of the Irish PEN anthology *Concord of Harps* in the April issue of *Poetry Ireland*, Marcus states:

> *Concord of Harps*, an anthology selected from the work of members of the Irish P.E.N. Clubs, is a pleasing and varied collection ... If, in this collection, one note more than another predominates, it is perhaps that which is known as 'Celtic Twilight', but there is frequent change of mood and theme. *Concord of Harps* does more than merely fulfil its intention of affording 'a glimpse of Irish poetic achievement in the present century'. (17:23–4)

Kavanagh's radically opposing opinion of this text apart, the style and tone, the entire ethos of this extract is from another discursive and political universe to *Kavanagh's Weekly*. It is altogether more nuanced and subtle, containing registers of possible irony and proviso for the initiated while still accessible to the amateur audience. It is undemanding and nonconfrontational, not overtly threatening its reader with the anxiety of acquiescence or demurral; there is, it appears, nothing more at stake here than a literary opinion. As Edward Said writes, however: 'Critics create not only the values by which art is judged and understood, but they embody in writing those processes and actual conditions in the *present* by means of which art and writing bear significance' (1991: 53). The 'processes and actual conditions' of this piece of critical writing – urbane, humanist, controlled, secured from any infection by the political – creates the conditions for a literary practice evincing precisely those same qualities. Furthermore, when placed in the context of modern Irish history, this tendency towards complete literary affiliation, couched in terms of essence ('Irish', 'Ireland') and difference ('universal', that which is non-Irish), represents a reaction against attempts by certain writers and intellectuals to displace earlier modes of decolonisation by locating apparently 'normal' cultural and political practices precisely in history, and showing how these discourses have arbitrarily emerged.[18] In attempting to step outside rather than negotiate the debates surrounding Ireland's recent colonial past, both *Irish Writing* and *Poetry Ireland* were accessories to that history, contributing to the maintenance of discursive structures whose persistence gave the lie to the prospect of a 'post-colonial' Ireland.

Like Kavanagh, many of the above individuals experienced the difficulty of trying to make a living through writing in post-Independence Ireland. Official apathy towards their work and the lack of a suitable readership fuelled the general shift towards critical discourse as the writer-critic transposed disappointment at his/her own fate into disappointment with post-Independence Ireland. Unlike *Kavanagh's Weekly*, however, all the literary/critical journals examined in this section demonstrate an inability to develop beyond the terms made available in earlier modes of decolonisation. In their discursive affiliations and critical approaches, each remained to some extent engaged in debates about the relationship between literature and nation which had their origins in the strategies of liberal and radical decolonisation. Although each occasionally displayed a capacity to displace, formally or conceptually, this relationship – whether it be an ironic self-interview by Temple Lane, or a nationalist history *ad absurdum* by Lord Dunsany, or a self-reflexive article on dramatic criticism by Valentin Iremonger[19] – each ultimately found the resonance from the revolutionary era still too strong to escape. Despite differences of affiliation and approach, this fate was by and large shared by those periodicals which engaged in specialist scholarly study.

The Professionals

There existed a genre of periodical publication in Ireland in the 1950s which, in terms of critical approach, political affiliation and implied readership, was far removed from either *Kavanagh's Weekly* or *The Bell*-influenced cluster of titles. This genre grew out of the specialist studies associated with research into all aspects of Irish experience, but especially its historical, literary and linguistic experience. The individuals engaging in these specialist studies were professional scholars who made their living from knowledge of lexicography, philology, mythology, topography, dialects, poetry, social history, genealogy and archaeology, as well as from the skills and techniques brought to bear on these fields: rigorous scholarship, analysis of primary sources, exegesis, translation, explication and interpretation. The sections to which these subjects and the journals they wrote were affiliated were relatively tiny specialist communities and institutions for the initiated – the universities, the Royal Irish Academy and the international guild of Celtic Studies.

The fate of the professional scholar in the postcolonial state is part of a larger question concerning the role of the subaltern intellectual in the decolonising process (Bhabha, 1994: 102–22; O'Dowd, 1985, 1991). Investment in serious, specialised cultural activity, combined with their guild- and later state-affiliation, meant that these scholars had little interest in questioning the dominant modes of Irish decolonisation or intervening in any way to expand the terms of the debate. To the contrary, their esoteric practices and narrow field of engagement meant that they were increasingly implicated in the reproduction of those discourses of uniqueness, essence and origin on which the received practices of liberal and radical decolonisation relied. At the same time, in their drive to conform to international standards of research, Ireland's professional scholars invested in a discourse which, in its implicit affirmation of the values of the élite cultures of Europe and America, was in fact fully implicated in colonialism (Lloyd, 1987; Said, 1985). Thus, decades after the supposed inception of the postcolonial era in modern Irish history, both the aims and the methods of scholarly discourse were still fully complicit with the maintenance of those strategies and techniques which guaranteed that the Republic of Ireland remained at a number of crucial levels a colonised country.

There is little of what during the 1950s passed for straightforward 'literary criticism' in these journals. However, given the emphasis on the correction and interpretation of texts there is a recognisable metadiscursive apparatus being deployed in which certain individuals, using a variety of specialist techniques and languages, are institutionally sanctioned to comment on a type of writing designated primary and valuable. Furthermore, with the juxtaposition of these two moments and two styles – the initial 'primary' writing and the subsequent scholarly writing – a principle of literary criticism is employed which, in the words of Tony Bennett, 'effects a specific ordering of the relations between texts, readers and practices of textual commentary' (1990: 196). It is not inappropriate, then, to analyse professional Irish scholarship in terms of its model of the relationship between culture, criticism and nation, and its intervention into the ongoing process of decolonisation.

As we have seen, there had been much amateur scholarly activity in Ireland since the late eighteenth century in the field of ancient and early modern Gaelic culture. These practices began to professionalise in the years after the Union under the influence of men such as George Petrie, John O'Donovan and Eugene O'Curry and the work they undertook for the Ordnance Survey between 1824 and 1841. The methodological techniques and archival work undertaken by these figures laid the basis for the modern scholarly discipline. This discipline, moreover, was highly influential on the emergence of cultural nationalism in the nineteenth century and contributed in a number of direct and indirect ways to the second Celtic Revival of the 1890s (Leerssen, 1996b: 102). The early generations of scholar-critics, however, had engaged in their work against an explicit political background: first Repeal, then Home Rule. Their work did not remain confined to a small initiated audience, but through the mediations of intellectuals and politicians – translators, as it were, from one discourse into others – found expression in the traditions and institutions of civil society (Cairns and Richards, 1988a: 42-57; Deane, 1985: 17–74, 1986: 60–89). Yeats and Synge, Connolly and Pearse, Griffith and de Valera – all found it useful, if not necessary, to engage with the findings of scholarly research when they attempted interventions in the Ireland of their day.

By the 1940s and 1950s, however, this political and civil context had been all but removed from research into Ireland's past, and an ethic, bordering on a cult, of professionalism had become dominant among practitioners. A number of prominent individuals – Osborn Bergin, Daniel A. Binchy, T.F. O'Rahilly, Eleanor Knott, Brian O'Cuiv, James Carney, Myles Dillon, David Greene and others – patrolled the borders of Celtic Studies, becoming a clique of 'textperts' with the institutional power to pronounce on the value of interventions into the discourse. At the same time, the direction

Celtic Studies was taking under their oligarchy was away from contemporary Ireland, and towards a narrowing and systematising of the discourse in which the political and cultural ideologies on which post-independence Ireland rested were considered off-limits. Liam O'Dowd writes that:

> as national boundaries become settled, intellectuals' explicit concern with cultural and national identity declines. Many traditional intellectuals assume specific functions in the new order. Specialist intellectuals come to have a much more prominent role in an intellectual stratum which is inclined to take cultural and national identity for granted. (1988: 10)

Techniques were refined and skills honed, but debates over the practical significance of such activities tended to disappear in the post-revolutionary environment. This knowledge-élite made a fetish out of method, employing specialised languages to communicate among themselves, as in this example from the journal *Ériu*, published in 1952:

> The poem is written in a form of *debide* metre, the scheme of which is: $7^{x} - 7^{x+1} - {}^{7x} - 7^{x+1}$. X is usually one but in fourteen verses is two and in one (v.8) we have $7^{1} - 7^{3} - 7^{1} - 7^{3}$ ... The second couplet of each verse has at least one internal rhyme and in many cases there are two. In v.22^{d} *osait* may have been so written to show the rhyme with *chosait*, though as *asait* is the usual form we may have to read *chasait:asait*. These rhymes are all perfect. (16:157–40)

A title of the Royal Irish Academy, *Ériu* was founded at the turn of the century by the prestigious Celticist Kuno Meyer as 'The Journal of the School of Irish Learning' and was 'Devoted to Irish Philology and Literature'. Its pre-Treaty leanings were towards radicalism, then, 'devoted' to the process of demarcating and instituting a field of experience – Learning, Philology, Literature – which would differentiate Ireland from other nations. However, although radical resistance (conceived in terms of the essential, authentic cultural experience of the decolonising community) had achieved its aims, *Ériu* persisted in refining the techniques of authentication and differentiation upon which colonial discourse relied.

The article from which the above extract is taken is titled 'A Middle-Irish Poem on the birth of Aedan Mac Gabrain and Brandub Mac Echach' by M.A. O'Brien, and it is typical of the approach and orientation of the majority of work produced in this kind of discourse. It proceeds by way of formal empirical exegesis and objective analysis of primary manuscript material; there is little or no indication of those discourses of evaluation, appreciation or reader response which characterise the critical approach of *Kavanagh's Weekly* or *The Bell*. There are likewise no signs of O'Faoláin's quasi-holistic pronouncements or Kavanagh's wish for traditional intellectual status. This reading is instead radically decontextualised, authorless and timeless to all intents and purposes. A discursive universe is implied in which writer and reader belong to an initiated community, but one which bears no relation to any community outside this particular universe. It implies, that is, affiliation to an artificially constructed section, one which speaks, thinks and lives apart from the issues which currently engage the larger national community.

One extreme direction this textual fetishism could take can be seen in the pages of occasional journals such as *The Irish Book*, *The Irish Bookman* and *The Irish Book Lover*. The latter was almost entirely written, edited, printed and published by Colm O'Lochlann at his own printing press, 'At The Sign of the Three Candles'. Here, individual texts were reviewed as in other periodicals, but the dominant evaluative criteria were the volume's production values, its availability and its price. Thus, when comparing a book on wood engraving with one on poetry, O'Lochlann writes:

> Of these two the Bewick one pleases me most as I think the use of the border on the Wordsworth pamphlet takes from its general appearance. The border is too near the edge, and in fact on my copy has been cropped by the cutter ... The edition is limited to 300 numbered copies, and I am sure that this will become a 'Collector's piece' in a short time. It is priced as 5s.; while the Bewick book only demands a modest 2s.6d. It also is limited to 500 copies. (32:6:146, September 1957)

Besides this kind of comment, the most typical critical statement to be found in the journal is: 'There is an immense amount of Irish interest in this book' (32:6:150, September 1957). Although amateur book-collecting has a different intellectual genealogy to professional scholarship and may seem an innocuous hobby irrelevant to the issue of decolonisation, as a discourse of writing about writing looking to intervene in Irish critical practice in the 1950s it is in fact fully implicated in questions about the relationship between literature, criticism and nation in postcolonial Ireland. Taken together, the above extracts represent a response to the text which is completely divorced from issues of analysis or explication or 'literary criticism' in any of its contemporary forms. The text becomes an object in and of itself, regardless of what or how it signifies. Despite the 'Irish interest' to be hunted down in each text, its value as a commodity is not influenced by questions about the relationship between culture and nation. It is the logical end of a regime of practices which, like specialist scholarship, values discourse for the material forms it takes rather than for its communicative or interventionary potential. In Kavanagh's terms, book-collecting in Ireland in the 1950s is diametrically opposed to 'life'; and the sort of critical commentary practised by those involved in book-collecting is a politically quietistic rejoinder to those contemporary writer-critics wishing to make criticism a powerful interventionary weapon in current debates.

Besides the tendency towards textual fetishism, the other main characteristic of these specialist studies was their historicist bias – historicist not only in the sense of a primary text's immediate context but also in the sense of the quasi-social-scientific method brought to bear on that text. Medieval and early modern Irish poetry, for example, was discussed not so much in terms of its 'literary' traits (however these might be construed) but as a primary source for historical research. The reviews of journals such as *Éigse* and *Irish Historical Studies* were often interchangeable in their scrupulous attention to matters of scholarly method and their blasé attitude towards literary effect. Thus, in the September 1948 issue of *Irish Historical Studies* P.J. Brophy reviews *Poems on the Butlers of Ormond, Cahir and Dunboyne (A.D.1400–1650)*, a volume edited by James Carney and published in Dublin by the Institute for Advanced Studies:

> The publication of a collection of poems in Irish on the Butlers is a matter of interest to students of Irish history ... With the exception of poems XV and XVI, both of which are devoted to the exploits of Black Tom, tenth earl of Ormond, the historian will find little new information in this collection. For the genealogist there are a few items of interest. Since the text presents many difficulties, it is a pity that this excellent edition of the Butler poems, furnished with a good introduction, copious historical and linguistic notes, several useful indexes and a glossary, does not contain an English translation. (6:139–41, September 1948)

Here, it is the 'edition' which is all important, questions about the 'introduction', 'notes', 'indexes' and 'glossary' taking an apparently natural precedence over the 'poems' themselves. Without these indicators of initiation, the scholar-critics can dismiss any attempts to recruit the text for other contemporary purposes. The methodological skills with which the 'historian' or the 'genealogist' confronts the poems are designed to discover and enunciate only certain things, while marginalising and ignoring others. They allow the scholarly subject to 'master' the text, and in so doing arrogate the right to say in what contexts it may be deployed. Any attempted intervention on the part of traditional writer-critics can be disbarred on the grounds of ineligibility and inaptitude. The text, in other words, is appropriated into a metadiscursive framework in which any immediate popular appeal is precluded, as are any contemporary political affiliations which might wish to harness such an appeal.

Such is true also of the more 'literary' journals such as *Éigse*, in which the following statements appear in an article entitled 'Two Poems of Invocation to Saint Gobnait' by Brian O'Cuiv in the 1952 collection:

> [A] In view of the frequent association of St Gobnait with cures of plague and disease, and especially with cures of smallpox, the present poems are of especial interest ... [B] I have transposed lines 56–58, [C] as the result is better metrically while giving equally good sense. The language of the poem is refreshingly simple and the meaning generally quite clear ... [D] The form of the poem is interesting, for it demonstrates Seán Ó Múrchádha's versatility in metrical composition. [E] Of the ten stanzas in the poem five are in *amhran* metre, but one of these differs from the others in having an extra (sixth) stress added at the end of each line. Of the remaining five stanzas four (sts.2–5) are in a loose form of *rannaiocht mhor*, while the fifth (st.6) is in loose *deibhi*. None of Seán Ó Múrchádha's poems published by Torna are in these syllabic metres. (letters added) (6:328–31)

In this extract appear all the characteristics discussed under the rubric of professional scholarship – historicist bias, discursive technologisation and methodological fetishism. In the first instance [A], the 'interest' and value of the poems lie not in their relevance to any identifiable section of the contemporary community or in any intrinsic 'literary' attributes, but in their ability to shed light on specific historical phenomena: the association of St Gobnait with the ability to cure diseases. Next [B], the author intervenes in the poetic process itself, using the skills of his profession literally to *create* the text which [C] is to be the subject of his scholarly pronouncements. The terms used to frame and evaluate the text ('good sense ... refreshingly simple ... quite clear') demonstrate the professional's bias towards objective, empirical discourse, and while the responses [E] are highly technical and esoteric, interpretations of the surface 'meaning' can be superficial and simplistic with the scholar getting so densely involved in questions of rhyme, rhythm and metre that the question of the relationship between 'meaning', form and context is ignored. The 'meaning' of the poem can be read off by those possessing suitable skills and qualifications, but there is no indication of the contexts, beyond the original, in which such a 'meaning' might signify; indeed, the 'meaning' found by the scholar dominates the other possible 'meanings' – in the sense of 'significations', the text's career beyond its author's intentions – which the text might have if allowed articulation outside the methodological restraints set by the scholarly system. Against the danger of this 'meaning' escaping into other spheres of discourse, the scholar produces a technical language which [D] simultaneously relocates the interest of the text in a specific historical figure (Seán Ó Múrchádha) while [E] marginalising any potential non-initiated intervention. The effect of the whole is to demarcate a realm of activity in terms of a range of techniques and attitudes, at the same time delimiting the affective range of this realm and arrogating to a small number of initiates the power to nominate the contexts in which it can be evoked. For the professional scholar-critics of postcolonial Ireland, it appeared no such context was suitable beyond their own tiny enclosed community.[20]

The politically quietistic stance of the professional scholars was compounded by the establishment of the Dublin Institute for Advanced Studies in 1940 when many of the subjects and organs involved in specialist studies became openly state-affiliated. The function of the Institute, according to the Act of Dáil Eireann under which it came into existence, was 'to provide facilities for the furtherance of advanced study and the conduct of research in specialised branches of knowledge, and for the publication of advanced study and research whether carried on under the auspices of the Institute or otherwise'.[21] Missing from this directive is a sense of the purpose behind all this activity or the context in which it will take place. The emphasis on specialisation airily ignores, in the words of Edward Said, 'the circumstances out of which all theory, system, and method ultimately derive' (1991: 26), and for all their historicist bias, postcolonial scholars seemed unwilling to contemplate the possibility that their own activities emerged from the same discursive matrices which enabled the practices of colonial power and anti-colonial resistance.

The Act mentions 'advanced study' and 'research in specialised branches of knowledge'; however, reading the dedications and acknowledgements, the citations and references of the numerous texts published in the following years under the auspices of the Institute, there is a sense that the activity has become an end in itself, that the community of scholars involved in Celtic/Gaelic/Irish Studies has become an extended and self-regulating family – a family prone to internecine squabbles, certainly, but one fundamentally united in its dedication to the maintenance of the family way of life. And this dedication ultimately comes to replace any sense of the reason behind all the 'advanced study' and 'specialised research'.[22]

The School of Celtic Studies was founded under the terms of the 1940 Act to promote research and regulate funding in the field. The School provided a forum where professors, lecturers and research assistants could circulate their work through a variety of forms including courses, seminars, conferences and journals. The name of the School's official journal was *Celtica*, and like the other journals examined in this section it is highly erudite, employing a variety of specialised technical languages written for an initiated audience. Contributions were mainly transcripts of medieval fragments, always highly annotated, sometimes with a glossary and an English translation, with the emphasis always on high scholarly standards. Longer pieces were commissioned and issued as separate volumes, and by the middle of the 1950s the School had four special series in publication: *Irish Franciscan Texts, Scriptores Latini Hiberniae, Mediaeval and Modern Irish Series* and *Mediaeval and Modern Welsh Series*.

Where the journals did occasionally threaten to break into something like traditional literary criticism was in the review section, but here again the dominant impression is of a small number of individuals talking among themselves, a professional élite employing esoteric codes and references to police the boundaries which demarcate both their intellectual realm and their livelihoods. Thus in the 1954 volume of *Celtica* one scholar (David Greene) reviews the work of another (Kenneth Jackson's *A Celtic Miscellany*), implicitly and explicitly invoking the professional community which forms their audience: 'Professor Jackson then goes on to warn the reader against the "Twilight" view of Celtic literature – surely unnecessary so many years after Synge renounced the "plumed yet skinny Shee" and took the vast majority of writers in the Celtic countries with him?' (2:210). The name of Synge functions here in a highly complex manner. The initial reference is not to the popular playwright hero of postcolonial mythology but to the serious Gaelic scholar commenting on what he believed to be a mistaken strand of thought in contemporary Celtic Studies. Only those aware of this aspect of Synge's career – 'the vast majority of writers in the Celtic countries' – will appreciate this reference. At the same time, the historical context in which Synge's comment was made is silenced, as is its potential political force – that is, its subversive, ironic attitude towards some aspects of the cultural politics of the Celtic Revival. 'Synge', therefore, is employed to shore up the contemporary discipline of Celtic Studies, while any practical political connotations associated with that name are marginalised. The profession denies a context outside itself – denies, that is, the existence of a section to which Synge's (serious) playful attitude towards Celtic discourse might be offered as an alternative to received decolonising discourses.

In all these journals and publications the same names keep cropping up, and always the predominant impression is one of a self-sufficient, self-perpetuating profession in which focus on the minutiae of 'Irish culture' has obscured the debate on the function of such a formulation and its role in the contemporary decolonising process. As one of the few critics who worked assiduously to maintain such a debate, Vivian Mercier, wrote in 1956:

> The gulf between scholarship and criticism seems even wider and deeper in Ireland than in other countries. Those who criticize don't know – those who know don't criticize. In other words, our critics are too unscholarly, our scholars too uncritical or too indifferent to the common reader. (86)

Such a limited and interdependent community is particularly prone to absorption by the state. Research, once it moves beyond the purview of the gentrified scholar, has to be funded. It is not likely that the authors or any of the individuals mentioned in this section could have rattled off a poem or an article for the more popular newspapers or journals to pay a month's rent. The thin society of post-independence Ireland could barely support the more accessible writer-critics of *Kavanagh's Weekly*, *Envoy* and *The Bell*; it was extremely unlikely, therefore, that pieces on 'Some Developments in the Imperative Mood' or 'Descriptoribus Hibernicis' would engage the journal-buying Irish public in enough numbers to keep the professional scholar in business.[23] The established educational institutions apart, therefore, the state becomes the only viable alternative. The foundation of the School of Celtic Studies in 1940 may in fact be seen as the culmination of the movement in modern Irish history from anti-hegemonic to hegemonic activity in the cultural-critical sphere – the point at which the nation's licensed intellectuals mortgaged their expertise to a state apparatus dependent on the maintenance of outmoded ideologies, thus contributing to the domination of the most reactionary elements in the postcolonial society.

Affiliated Periodicals

Another kind of periodical available in Ireland during this period shared some of the aims and techniques of the titles already examined but was affiliated to specific institutions or sections, most notably the university, the church and the intelligentsia. The subject matter, method and implied readership of journals such as *The Dublin Magazine*, *Studies* and *Hermathena* were often just as focused and esoteric as those of contemporary Celtic/Gaelic/Irish Studies, although they lacked the austere professional ethos of those disciplines. At the same time, a certain amateur air which it shares with *The Bell*-inspired publications surrounds much of the material contributed to these journals, although there is not much evidence of the holistic pretensions of writer-critics such as O'Faoláin and Kavanagh. These journals, in other words, existed as a sort of half-way house between opposing models of what a 'national' literature and a 'national' criticism should be in postcolonial Ireland. They were erudite, scholarly and often just as closed off to a more general readership as were the specialist publications of the School of Celtic Studies, yet their sectional affiliations meant that they were not entirely divorced from the political considerations of civil society.

This ambiguity of orientation, however, rather than creating the possibility for a displacement of the established discourses of postcolonial Ireland, in fact aided their consolidation. This is because the sections to which these journals and their contributors were affiliated were heavily implicated in the received lines of thought along which colonialist and decolonising discourse had continued to operate since the colonial era. Although method and affiliation might often be at odds in this material, there was nothing active or self-conscious in such a juxtaposition, but rather a determined effort to map one onto the other in the cause of some essential, authentic, completed Irishness. The Catholicism of the Maynooth-sponsored publications, the classicism and residual Anglo-Irish sensibility of Trinity College, even the cosmopolitanism of those who considered themselves members of the national intelligentsia – all these worked within the limits set down in the cultural nationalist discourses of the pre-revolutionary period, that time when anti-colonial resistance had oscillated between liberal equality with, and radical difference from, the imperial subject. The postcolonial nation imagined in these journals had little to do with the bohemian artists of Grafton Street or the citizenry of the rural village, or even the artificial political divide between certain parts of the island; rather, it was an abstract concept which thirty years on was still being articulated in the accents of Ireland's colonial and revolutionary history.

These periodicals, in fact, were all implicated in the coming to hegemony of the nationalist bourgeoisie, that section of the community which effectively hijacks the revolution and comes to dominate, as blatantly as the ousted colonial regime, the post-independence nation (Fanon, 1986, 1967). This hegemonic formation worked for a cultural ideology in which 'national' and 'normal' would be synonymous (Corkery, 1931: 3). At the same time, as a *hegemonic* (as opposed to a dominating) discourse, it provided for various dissensions from that ideology which, whatever their orientation – religious, classical, cosmopolitan – would still be implicated in the maintenance of that central ideological connection between nationality and normality, between culture and nation. Thus, while certain contradictions between critical approach and sectional affiliation raised the possibility of some kind of perspective on the discursive categories of postcolonial Ireland, the engagement with received models of critical/creative activity militated against any such perspective. The 'literary criticism' in which these journals engage bears little relation to the more wide-ranging 'cultural criticism' practised by many of the contributors to *The Bell*, *Envoy* and *Irish Writing*, or to the methodological fastidiousness of *Celtica*, *Ériu* and *Éigse*, while in tone and style it is worlds away from *Kavanagh's Weekly*. Like these publications, however, journals such as *University Review*, *Threshold* and *The Furrow* were fully implicated in the ongoing issues of decolonisation. Like the majority of the journals of the 1950s, in fact, these latter titles were complicit with the essentialist ideologies deployed by the nationalist bourgeoisie to dominate the postcolonial state.

The Dublin Magazine: A Quarterly Review of Literature, Science and Art was founded in 1923 by the Dublin-born poet and all-round man-of-letters Seamus O'Sullivan, a man who numbered most of the major and minor figures of the Literary Revival among his acquaintances. With the animosity of the Civil War still tangible and political events still dominating the country's attention, 1923 might have seemed an inauspicious time for the foundation of such a journal. Nevertheless, 'it became one of the most influential ever published in Ireland, and over the next thirty years almost every Irish writer of merit contributed to it' (Brady and Cleeve, 1985: 200). A contemporary review in the *New York Herald Tribune* agreed: 'The reader will find in this journal all that is best of Irish thought, and its list of contributors includes the names of all the best-known Irish writers.'[24] *The Dublin Magazine* could also boast an impressive list of international contributors, and every so often it published a list of these names as if to confirm its high domestic and international reputation. The list in the July–September issue of 1952 contains around sixty names, many of whom are from Great Britain, Europe and the United States.

However, unlike *Envoy*, for example (the contemporary Irish journal to which it is closest in orientation), the critical and creative contributions favoured by O'Sullivan tended to be from well-established, commercially safe names engaging in kinds and ranges of discourse which would have been familiar to the journal's readership. Unlike *Envoy*, that is, *The Dublin Magazine* is far from being an avant-garde 'little magazine', either in size or editorial approach. Rather, its impressive intellectual credentials and its self-conscious European outlook mean that *The Dublin Magazine* aspired to join a wider cultural network in which contributors and readers could engage, regardless of nationality. Together, these contributors and readers represent a sectional affiliation characterised by intellectual, educational and class, but not necessarily national, determinants. Nationality, if it must be discussed at all, is an incidental factor in questions of literature, and excessive reliance on it as an aesthetic criterion represents a failure in education and taste. *The Dublin Magazine*, in fact, represents a strand of postcolonial thought which, by its self-conscious affiliation to a section not identifiable in specifically national terms, hopes to out-manoeuvre the debilitating effects of residual colonialism. However, as remarked above, this attempt to get 'beyond' the parameters of nationalist/colonialist discourse in fact reinforces the structures – inside/outside, Irish/non-Irish – which feed and sustain such discourse.

This can be seen in the critical material included in *The Dublin Magazine*. For example, the English academic and poet Donald Davie in an article published in the January–March issue of 1952 wrote that:

> [The controversy over Synge] is a matter which can be decided only by Irishmen, in which the word of a foreigner such as myself can carry no weight ... Synge's volume is a challenge not only to the Irish poets of his time, but to everyone who tries to write poetry in English at any time. Synge wants to shock the inhumanly exalted poetry of his time by writing poems of all too human degradation; but only in hopes that between the two extremes poetry may come to rest in a central area of human interest and compassion ... He remains one of the very few poets,

> writing in English since the end of the eighteenth century, who have talked sense about the question of diction in poetry.[25]

The question of nationality is dismissed in favour of what this critic obviously considers the more engaging issue 'of human interest and compassion'. 'Poetic diction' is a suitably universal topic – relevant to 'everyone who tries to write poetry in English at any time' – to engage not only 'Irishmen' (when they can drag themselves away from the less important controversy over Synge's national credentials) but the great international community of poets (and, by extension, readers and commentators) addressing this far more interesting question. The fact that Synge wrote in English is incidental to his achievement. There is no reference to the arguments which in Ireland had surrounded his work since his death – arguments concerning cultural imperialism and the nationalist drive towards cultural confidence through literary and linguistic independence. Such debates are a sign of critical immaturity and have no place in the pages of such a sophisticated publication.

At the same time, in referring to himself as a 'foreigner' Davie raises the whole issue of nationality and appropriate behaviour, seemingly contradicting the universalist impulse of the whole piece. This is not an oversight on the part of such an astute professional academic, but a symptom of this journal's inability to free itself from the restraints of colonialist discourse. The critic, in other words, retrieves Synge from the margins (or the provinces) of English literary debate to a position where the real 'challenge' and significance of his work can be discussed. However, there cannot be equality without difference, and Davie's article mirrors the typical structure of colonialist discourse with its initial reference to that which is to be dominated and marginalised (the Irish controversy over Synge), followed by a universal, salvational discourse which draws its discursive power from that which it perceives itself not to be. This critical manoeuvre, however, rather than 'raising' Irish experience up to a universal (that is, English) level, recalls the liberal mode of decolonisation in which a formerly marginalised subject seeks equality with the metropolitan power, only to reconfirm, at the very moment of its 'success', the hierarchies on which the established structure of power and knowledge was founded. To the degree that Synge's work is universal (non-Irish) and is discussed in universal (non-Irish) terms, then, Irish experience is simultaneously confirmed and contained.

This can be seen again in the issue for April–June of the same year in which the work of George Moore is discussed by a reviewer, in terms which simultaneously deny and confirm a unique Irish experience:

> His prose, always fastidious, is at the same time astonishingly simple so that his novels are often enjoyed by uneducated as well as sophisticated readers ... His novels remain remarkable because in them his integrity and his humanity are expressed and expressed with a crystal precision of style. Considering this, is it surprising that even among those many Irish writers who have enriched the literature of both the past and present centuries, he should still hold a distinctive place? (27–31)

This extract is typical of the polished style and cultured tone in which most of the criticism in *The Dublin Magazine* is written. The response is slightly more than 'appreciation' but somewhat less than 'analysis', with loose terms such as 'integrity' and 'humanity' balanced by pseudo-critical terms such as 'fastidious' and 'crystal precision of style'. In this way the article and the journal signal their affiliation to an enlightened, educated community at home and abroad. The reference to the uneducated/sophisticated opposition also betrays the critic's orientation, implying two ways of reading and responding to a text. These ways, however, are not equally valid; rather, one (lesser) response is announced in terms of another (greater) response, thereby constructing an unequal relationship based on access to discourses of power and knowledge. And this structured-in critical difference is repeated when it comes to questions of nationality, for despite the ability of 'Irish writers' to gain access to the universal phenomenon of 'literature', the traditionally unequal national categories of 'Irish' and 'non-Irish' are not exhausted by such a formulation; rather, they are sustained and reconfirmed by it. The rhetorical question which brings the piece to a close attempts to disguise the discrepancy between a canon of 'Irish writers who have enriched the literature of the past and present centuries' and those other Irish writers – many? most? – who have not been sufficiently educated or intellectually endowed to raise themselves above their local experience to engage in issues of non-Irish, universal import. The drive for equality only serves to reconfirm the existence of difference.

The large number of church-affiliated journals published under the auspices of Maynooth, the country's centre for religious study, by and large did not share *The Dublin Magazine*'s anxieties over equality. As organs of what they believed to be the single most important characteristic of southern Irish identity – Roman Catholicism – publications such as *Hibernicum*, *Christus Rex* and *The Furrow* were fully and self-consciously implicated in the continuing dominance of modes of thought which had been fostered under the impetus of radical decolonisation, emphasising those factors which differentiated authentic Irish from non-Irish experience.[26] If that experience connoted highly sophisticated intellectual intercourse – as it did, for example, in the pages of the Jesuit publication, *Studies: An Irish Quarterly Review of Letters, Philosophy and Science* – such discourse was always ultimately subservient to what amounted to the final word (religion) in every Irish/non-Irish encounter. And as long as authentic Irish experience could be identified so readily and so finally in terms of this one experience, the continuing restrictive influence of colonialist/decolonising discourse was guaranteed.

What complicated the situation, however, was the fact that this unique Irish experience was identified by individuals employing techniques and skills which appeared to emanate from a discourse of free intellectual inquiry. *Studies* did not shy away from the potentially embarrassing questions raised by modern art and literature; rather, it confronted these questions on their own terms but resolved them by employing arguments and tactics emanating from theological discourse. Thus, in the issue for March 1952, M. Bodkin, S.J., praises a novel on the life of Joan of Arc ('A Halo on the Novel', 77–82) but recommends the actual records of her trial because 'they give us St Joan herself'; Ludwig Bieler concludes a highly technical and allusive article on 'The Nature and Meaning of Language' (83–90) with the statement, 'When pursued to its logical conclusion, philosophy ends up as anthropology, anthropology as theology'; while T.P. Dunning is embarrassed to report that he enjoyed a book on 'The Meaning of Beauty' (112–15) even though the writer rejects the authority of St Thomas Aquinas.

It is in the same issue that a long review article of Benedict Kiely's *Modern Irish Fiction* by the established man-of-letters Francis MacManus (who was at the time also Director of Talks and Features on Irish radio) reveals the ideological implications of church-affiliated literary criticism in Ireland at this time. In this article on 'Modern Irish Fiction' (121–3), MacManus agrees with Kiely that the crux of Irish literature, although usually expressed in cultural and political terms, is fundamentally the religious crux of acceptance and rejection: the sacred pleasure of succumbing to that which is familiar or the secular pleasure of embracing the unknown. MacManus puts it this way:

> Do I accept the material that is to my hand and that lies dormant inside me as unformulated and artistically unexpressed experience with all its

> burden of history and tradition, society and religion, its implications and contradictions and echoes? Or do I for the most part reject and manipulate the material satirically in the spirit of a reformer or with the primitive zeal of the iconoclast? ... The theme is universal in criticism as in creativeness. Across Mr Kiely's period of study falls the influence of two writers, one a Corkman, the other a Dubliner; one seldom saluted for his worth, the other continually and tiresomely celebrated, and the writers are Daniel Corkery and James Joyce. (122)

Such a clearly delineated opposition, personified in two famous literary figures, can only add to the either/or, same/different system on which residual colonialist discourse depends. The pejorative tone regarding the latter option is a hint as to where the reviewer's (and, presumably the editor's) sympathies lie. The article, the journal and the 'useful, serenely discursive, and witty book' under review (121), all are recruited on one side of a spurious divide which, in a religious register, mirrors and partially constitutes the limits of Irish and non-Irish experience. In this way, the latitudinarian intellectual rhetoric of much postcolonial religious discourse could be transformed into what Liam O'Dowd calls 'an exceedingly narrow and authoritarian social ideology for popular consumption' (1988: 13).

The Dublin Magazine and *Studies* were affiliated to imaginary cosmopolitan and domestic audiences which reflected and shared their tendencies towards certain characteristics of class, education and religion. Although its general orientation was similar, there was nothing imaginary about the readership implied and sued by *Hermathena: A Series of Papers on Literature, Science and Philosophy by Members of Trinity College, Dublin*, which, as its fly leaf announced, accepted articles 'only from Graduates (honorary Graduates included) of the University'. Seldom could an Irish periodical have been produced by and for such a precisely delineated section. In style, format and content, *Hermathena* made no secret of its affiliation to a tiny, highly educated community characterised by an attitude of general indifference towards matters of modern national identity and decolonisation, something reflected in its dual publication address – Dublin and London. This attitude, however, was misleading, as both the journal and the institution to which it was tied were in fact fully implicated in the (anti-) colonialist discourses which continued to dominate the national life decades after the official end of colonialism.

For Trinity in general and for *Hermathena* in particular it seemed as if the period in Irish history between 1890 and 1922 had never occurred. The Catholic, Romantic, Nationalist Ireland which had come into existence after 1922 bore no relation to the Protestant, Classical, Unionist Anglo-Ireland of Trinity's heyday in the eighteenth and nineteenth centuries. The university's response to this unfortunate situation was to develop a siege mentality and carry on regardless, pretending that the cultural work carried on inside the walls was not affected by the political developments outside.[27] The critical orientation of a journal such as *Hermathena*, therefore, remained much as it would have been a hundred years before – amateur, understated, decorous. Its subject matter consisted largely of articles on, and reviews of, classical and neo-classical literature and philosophy. When Irish-related material was included, it generally retained that air of anthropological curiosity which the early Anglo-Irish scholars had brought to the fashionable hobby of Gaelic antiquarianism. Employing these tactics, the tiny section involved with the journal hoped to offset the worst effects of the tragedy which had overtaken modern (Anglo-) Ireland. Far from undermining the power of the nationalist bourgeoisie now ensconced in Dublin Castle, *Hermathena*'s haughty, disengaged stance was in fact part of the process whereby liberal and radical decolonisation continued to set the limits on contemporary cultural/critical behaviour.

The issue for May 1952 (which coincided with the brief life of *Kavanagh's Weekly*) served up the usual diet of classics and philosophy, contained in such articles as 'The Curse of the Alkmaiondai – Part II' by G.W. Williams (3–21); 'Favilla, with Special Reference to Propertius, 1.9. 17–18' by W.J.N. Rudd (30–3); and 'Whitehead's Irrationalism' by R.R. Hartford (89–93). There were also three articles relating in some degree to Ireland: 'A Note on St Patrick in Gaul' by E.A. Thompson (22–9); 'The Book of Kells and the Gospels of Lindisfarne – a Comparison' by A.A. Luce (61–74); and 'Somerville and Ross' by Sir Patrick Coghill (47–60). (There also appeared an article entitled 'Studies in the Characterization of Ulysses – IV' by W.B. Stanford, a Trinity scholar who carved a unique niche for himself in modern Irish letters by identifying echoes of the classical Ulysses in Joyce's novel.) All the articles and reviews, however, are written with the same self-assured tone and relaxed style that comes with the confidence of belonging to an educationally privileged community.

Coghill, for example, was not a professional scholar or academic. Indeed, apart from his Trinity education, his only qualification for delivering a lecture on Somerville and Ross (of which the *Hermathena* article is a transposition) appears to be his blood relation to the two cousins.[28] Given his education and connections, however, he still appears confident about this intervention into the specialised area of modern Irish literature, and about his own ability to put together an engaging piece for a like-minded readership. 'My plan', he writes at the outset, 'is to try to give you a picture of the lives and backgrounds of these two gifted cousins, a reconstruction of their methods of work and some tentative speculations on the mystery of their unique collaboration' (47). And that, indeed, is what the piece does, giving the reader a biographical account of the lives of Somerville and Ross followed by some first-hand reports on their writing practice, concluding with a judgement on the particular traits which each cousin brought to the collaboration. There is nothing technical or specialised about the discourse. The reader is invited to participate in a refined process of discussion, appreciation and evaluation, implying a shared code of educational, aesthetic and social sensibilities:

> What does stand out is the extraordinary union, fusion rather, of two minds and two natures; as closely one and yet as different as are the warp and woof of some fine Persian carpet. But when all is said and done, genius defies analysis. We must accept it gratefully and leave it at that ... It is a matter of astonishment, admiration and envy that two girls bred in the wilds of West Cork and Galway could become master writers without, apparently, a long, gruelling and discouraging apprenticeship ... Therefore, not only this island of ours, but the greater Ireland scattered over either half of the world, will confess, and gladly, that for no women's brow could our academic wreath more worthily be woven. (59–60)

There is an amateur, anti-theoretical (even anti-intellectual) ethos which pervades this article and much of the material included in *Hermathena*. The writer admits that this is not to be a 'detailed critique'; in fact, he does not appear to possess the skills and languages necessary for such an undertaking. Instead, Coghill falls back on rhetoric, employing techniques of analogy ('some fine Persian carpet'), aphorism ('genius defies analysis') and fine writing ('for no women's brow could our academic wreath more worthily be woven'). These are skills and traits which the educated individual – and especially a Trinity graduate – would be expected to possess. They speak not of the specialist techniques of a professional scholar like Brian O'Cuiv, or the holistic cultural criticism of Seán O'Faoláin, or the deliberate hit-and-miss style of an impoverished writer-critic like Patrick Kavanagh. They speak, rather, of refinement and knowledge acquired through privileged education and leisure.

Also, there is the question of whom the author is referring to when he writes of 'this island of ours'. The surprise effect ('astonishment, admiration and envy') which for two and a half centuries had been one of the main weapons in the Anglo-Irish armoury – the unexpected emergence of refinement and talent from a wild, unlikely milieu – is wheeled out and fired in the face of both Irish-Irish disbelief and English disinterest. With its biographical narrative of the lives of Somerville and Ross and the natural way they blend into their Irish surroundings, the article is in fact a celebration of Anglo-Irish achievement and a validation of the Anglo-Irish experience. 'Ours' is both a proud claim for recognition and a desperate plea for survival, an attitude which sums up the overall orientation of *Hermathena* itself.

By employing the techniques of classical rhetoric for what amounts to an apology for Anglo-Irish life, Coghill is announcing his affiliation to a residual, educated, 'naturally' enabled section which, despite the pretence to indifference, cannot disguise its resentment at the hijacking of the national name by a larger, more powerful, but less deserving, section. The attitude of business-as-usual adopted by Trinity and *Hermathena* in the years after independence, however, did nothing to undo the colonialist discourses which gave rise to their marginalisation from modern Irish life. Rather, the dogged adherence to certain models of Irishness and Anglo-Irishness (models which had their roots in the colonial period) perpetuated the structures of thought which had led to the contemporary situation, arresting the process of decolonisation just at the moment when it needed to gain some perspective upon its own possibility.

Journals similar in orientation to the three examined above – such as the National University's *University Review*, Maynooth's *The Furrow* and the Belfast Lyric Players' *Threshold* – although each possessing a unique set of publication and circulation circumstances, could expect to face many of the same problems. Each was engaged in serious critical discourse which impinged on individual and social discourses of decolonisation. Each, however, continued to construe a relationship between Irish literature and politics which ensured the persistence of received structures of critical (and thus political) thought. In analysing these sets of periodicals, a range of critical modes – reviewing, biography, exegesis, appreciation and so on – are encountered. Such modes, moreover, were affiliated to a range of political positions. These complex cultural/political affiliations operated somewhere on a continuum between complete commitment to, and complete alienation from, post-independence Ireland. As will have been noted, the period is remarkably fluid, with various strategies and discourses employed in various contexts, often for antithetical ends. A sense of exhaustion pervades much of the periodical literature of the era, with the same old spurious choices between native loyalty and cosmopolitanism, between a national culture and a universal art, still setting the terms of the debate. There is much evidence of a destructive (indeed, of a deconstructive) impulse, of the need, as Seán O'Faoláin wrote in an angry and frustrated 'Signing Off' editorial in *The Bell* in April 1946, to clear away unsentimentally the brambles from contemporary Irish life. There is less sign of a constructive principle of regeneration and growth, of the positive displacement of received discourses which would herald a postcolonial, as opposed to a post-revolutionary or post-Independence, age.

The tension caused by this negative/positive hermeneutic helps to account for the rise during this period of literary criticism as a subject in itself rather than as a mere transparent point of engagement with the primary text. Literary criticism became the site of struggles over competing narratives of Irish history and different modes of postcolonial experience rather than a neutral discursive mechanism – becomes, that is, the location of 'a complex series of negotiations and compromises between varied interest groups' in postcolonial Ireland (Bennett, 1990: 237). The critical question in the years after 1948 becomes, not so much: what can this text tell us about Ireland?, but, in contemporary Irish cultural debate, who can say what about whom, and how?

Terence Brown refers to the writers of the 1950s as the 'tragic generation' (1985: 237) of modern Irish letters, and this is borne out in the material examined in this chapter. Humour, that comic double-vision which is the *sine qua non* of Joycean aesthetics, is the one element that is in relatively short supply in Irish cultural debate of this time.[29] Apart from Patrick Kavanagh's attempt to imagine a self-reflexive encounter between parish and empire, and his general iconoclastic attitude towards post-revolutionary Ireland, it is difficult to discern any sustained effort to think beyond residual decolonising discourses in the journals of this period. The various literary criticisms practised all rehearsed and perpetuated, to a greater or lesser extent, the received relationships between nation, literature and criticism, and the subject categories on which these relationships relied. 'Irishness' – whether construed in holistic, specialist or philosophical terms – was still the ultimate aesthetic criterion, and there was as yet no indication of the critical decolonisation of the text which would pave the way for a general cultural and political decolonisation of the mind.

6 The Scope of Literary Criticism

The periodicals examined in the previous chapter provided a space for a traditional form of literary criticism in Ireland in the 1950s, a space in which the principle of metadiscursive commentary could be brought to bear on certain kinds of writing valued within that society for the potential effects on its members. In Chapter 7 we shall be going on to analyse one other such form – the book. Literary criticism, however, was not confined to such traditional spaces; the discourse of commentary permeated contemporary Irish society, making itself available in many forms and many spaces not traditionally associated with the practices of textual commentary. When investigating the ways in which a critical principle was employed to organise the relations between texts, readers and society in the context of Irish literary criticism of the 1950s, therefore, it is necessary to talk of *criticisms* and *functions* rather than in the singular mode which has tended to dominate this problematic in recent years. Moreover, one must remain alive to the issue of 'what roles might be performed by different types of critical practice given the varied institutional domains, and their varied publics, in which such practices are operative'.[1]

This is especially true of postcolonial Ireland – north and south – in the 1950s where traditional critical discourse, by virtue of its established symbiotic relations with creative discourse, suffered in the general downgrading of cultural activity and thus had to relocate itself in various institutional spaces not immediately amenable to metadiscursive intervention. At least since the debates of the 1890s and 1900s, the written word had been taken very seriously at every social and political level throughout Ireland, and the capacity of literary critics to intervene in the narrative of decolonisation was widely recognised and valued. It should not be surprising, therefore, to discover the effects of the battle for control of the written word not only in the traditional locations of critical activity (such as the journal and the book) but in the multitude of forms and spaces throughout Irish society in the 1950s where 'the relations between texts, readers and practices of textual commentary' (Bennett, 1990: 197) were negotiated. Some of the forms and spaces in which such a critical negotiation could be analysed, and some of the many issues surrounding them, include:

Belles-lettres: was there still an audience for this kind of discourse, represented for example in the writings of Patrick Campbell, Oliver St John Gogarty or Lord Dunsany, in a society which had become highly suspicious of all 'intellectual', but especially non-national, cultural activity?

Theory: how did new psychological, political and cultural models of the ways in which literature functions impinge on traditional notions of the relationship between creativity and nation?

Scholarship: what models of literary and historical scholarship held sway, what archives did such models take as their constituency and what means were used to disseminate the findings of scholarly research to a wider readership?

Book retailing: who was buying, selling, printing and reading what kinds of 'literature' in Ireland in the 1950s, and for what reasons?

Imaginative writing: how were principles of commentary incorporated into creative or 'primary' discourse? To what extent did criticism affect the style and/or subject matter of novelists, poets and playwrights?

Official sources and reference works: we have already acknowledged the importance of Bodkin's *Report on the Arts in Ireland* (1949) and Whitaker's *Programme for Economic Expansion* (1958); we might also examine documents such as the census returns of 1946, 1951 and 1961 on issues such as literacy, disposable income and profession, or the *Report of the Commission on Emigration and Other Population Problems, 1948–1954* (1956) on the relationship between emigration and cultural identity.

Philosophy and religion: from the esoteric investigations of the universities to the populist strictures of a publication such as the *Irish Rosary*, how did issues of belief and truth relate to questions about the relationship between literature, the nation and critical commentary?

Historiography: what models of literary effectivity did different kinds of historians – social, political, cultural, economic – employ in interpreting the role of writing in modern Irish history?

Prizes, awards, scholarships: what criteria determined success in competitions such as the Adam Prize for Poetry, the Guinness Poetry Award, the AE Memorial Prize, the Harmsworth Award, the Casement Award, the Foyle Award, and in bids for the newly inaugurated Fulbright scholarships? (Whelan, 1996).

Radio: what effect did this important new medium have on traditional notions of the function of literature and the function of criticism?

Theatre productions: who made the decisions about what kind of plays were performed in the theatres around the country? How were these plays directed and acted, and what kind of audiences did they attract?

Re-issues: major editions of Moore, O'Casey, O'Connor, O'Flaherty, Shaw, Stephens, Swift, Wilde and Yeats, as well as many minor figures, appeared in the period under investigation; what factors (reputation, stature, demand?) influenced publishing houses to collect and/or re-issue literary material and how did this material relate to contemporary literary production?

Space: in what ways were the general principles of travel, environment and geography employed to confirm or problematise received notions of a national literature?

This is just a selection of the many forms and institutional spaces in which a critical apparatus could operate in Irish society during the 1950s. The narrative of decolonisation is just as available for confirmation or displacement in these discourses as it is in the more traditional critical forms. In this chapter I wish to analyse three such practices, looking at the ways in which the received relationship between literature and nation (and the forms of identity feeding into and out of such a relationship) was negotiated in ways not traditionally amenable to such an analysis. As in the previous chapter, we shall find a range of political affiliations and critical approaches cross-fertilised to produce and naturalise a range of postcolonial opinions. One might have hoped that here, in the marginalised and often covert critical spaces of the postcolonial state, there existed the best opportunity for decolonising subjects to imagine forms of critical engagement which might evade the insidious structure of received models. As with the investigation into periodical discourse, however, we shall find that these non-standard criticisms produced no sustained attempt to undermine the effects of what many in Ireland in the 1950s perceived to be an obsolete model of cultural activity. Rather, discourses of censorship, higher education and anthologising functioned (often against the conscious intention of critical subjects) to shore up an entrenched cultural nationalist ethos, thus helping to consolidate the narrative of radical decolonisation which had dominated – proactively or reactively – the whole island since 1922.

A Note on Censorship

Censorship is an extreme form of literary criticism, and it was an extreme form of censorship that was endorsed by the various Free State, Éire and Republic of Ireland administrations in the decades following the first Censorship of Publications Act in 1929. The sanctions and prescriptions which operate at a metaphorical level in traditional literary critical discourse are realised in the practice of censorship: *that* kind of literature is not compatible with this kind of nation or state, but *this* kind of literature is.[2] Thus, censorship both mirrors and helps to construct the identity of the nation it serves, employing evaluative and theoretical criteria in its selection of the cultural material to which such an identity can be exposed without fear of contamination or erosion.

The original intention to curtail the amount of 'indecent' and 'immoral' material entering the country may be seen as a manifestation of the fear, which the Free State shared with many parts of the Western world, of the impact on traditional values and practices of modern mass culture.[3] In the south, however, this fear was over-determined by the identitarian struggles peculiar to a decolonising nation-state. Despite certain safeguards incorporated into the Act of 1929, and one clause (Section 6, sub-section 3a) which specifically required the Censorship Board to have regard to 'the literary, artistic, scientific or historic merit or importance and the general tenor of the book or the particular edition of a book' (quoted in Adams, 1968: 238), the definition of 'indecent and immoral' soon became a catch-all for those many practices, interests and characteristics condemned as repugnant to the Irish identity as defined by the newly hegemonic nationalist bourgeoisie. Moreover, as Terence Brown has argued, it was over this specific issue 'that the interests of those who sought censorship from moralistic impulses alone and the interests of those like the Irish Irelanders, who desired cultural protectionism, met and often overlapped' (1985: 70). Which is to say, church-affiliated agents interested in protecting Ireland from developments in the area of family planning and what was considered a sex-obsessed modern literature found common cause with that section of the community concerned to define authentic national experience in terms of organic continuity and predominantly rural codes and practices. Together they produced an official 'Irish identity' which the censorship duly came to support – a narrow, insular and triumphalist category which took to extremes those psycho-social discourses typical of the radical mode of decolonisation.

As many commentators have noted, post-Treaty censorship represented in many ways a betrayal of the 'freedom' for which so much Irish blood and ink had recently been sacrificed (Blanchard, 1955; Brown, 1985; Ó Drisceoil, 1996). The hypocrisy with which the new state could define 'immorality' solely in sexual terms while ignoring the 'immorality' of enforced rural emigration and urban squalor led the historian Joseph Lee to opine that 'censorship,

Irish style, suitably symbolised the impoverishment of spirit and the barrenness of mind of the risen bourgeoisie, touting for respectability' (1989: 159), while Brown observes that 'none of the Irish-Ireland movement rose to decry censorship as a reactionary offence to the revolutionary humanism that had originally generated their movement' (1985: 73). Elsewhere Brown stresses censorship's ideological basis:

> Much more than a law to suppress the grosser forms of pornography, [censorship] had been revealed to be a legal instrument that could be used to protect Ireland from contamination by 'alien' influences and by Irish writers who could not accept the dominant moral and social consensus. The political censorship of the war years can accordingly be understood as a further attempt by the political class in Ireland to use laws restricting freedom of expression for ideological purposes. In both the literary and political censorship, national identity was at issue. Both involved notions of Irish exceptionalism, which presumed spiritual and moral superiority to other nations at its heart. (1997: 46)

At the beginning of the period in question, Éire (soon to be the Republic of Ireland) had recently entered what Michael Adams calls 'a distinct second phase of censorship' (1968: 115). A second 'Censorship of Publications Act' attempting to redress the abuses to which the application of the original legislation was prone had been passed in 1946 with the most important innovation being the introduction of an Appeal Board – 'a more enlightened group of citizens' than the Censorship Board itself, as Brian Inglis wrote at the time (1952b: 726) – where questionable decisions could be re-analysed. This action was partly in response to the increasingly embarrassing situation brought about under the previous dispensation, in which much of the work of many living Irish writers could not be read by their fellow countrymen and women; in which the Vatican's *Index Librorum Prohibitorum* (denying access to writers such as Zola, France, Gide, Sartre, Voltaire, Kant, Bergson, Gibbon, Paine, Hugo and Balzac) operated in a semi-official capacity at local parish level; and in which the Censorship Board's periodical publication of banned works read, as one contemporary commentator put it, like 'a concise guide to modern literature' (Blanchard, 1955: 72). This situation was unlikely to increase the country's international credibility, while it did nothing to appease the suspicions of those in the north who continued to see partition as a defence against 'Rome Rule'. At the same time, the country could hardly reap (in the shape of the tourist and culture industries) the potential economic benefits of the Irish contribution to modern literature if most of its famous writers were legislatively linked with pornography and immorality.[4] Unlike education, marriage, divorce and birth control, censorship was an area where the nationalist bourgeoisie seemed willing to negotiate on questions of Irish identity. After all, as contemporary opponents of censorship pointed out, there was room for interpretation (and thus for manoeuvre) on what constituted 'literature' and 'art' as opposed to obscenity, immorality or indecency.[5] The Act of 1946 was an attempt to acknowledge such arguments and thus make a gesture towards a more open, less insular society.

However, rather than relaxing censorship practices, the Censorship of Publications Act of 1946 seemed to exacerbate the situation, or rather, the interpretation and execution of the Act did. The number of banned books rose steadily, peaking in 1954 at 1,034 prohibitions out of 1,217 books examined, or six times the figure for 1949 (Adams, 1968: 119). It is this ratio, rather than the actual number itself, which is perhaps the best indication of the reactionary direction taken by official censorship in the years after 1946. Publications such as *The Bell* and the *Irish Times* attempted to keep the issue of censorship alive in Irish political and cultural debate, questioning at various times the concept itself, the Irish Censorship Board's interpretation of the legislation as well as the various forms of unofficial censorship practised throughout the country. What these developments show is that the religious and political sections which had dominated the trajectory of Irish decolonisation since 1922 were perceived to be under threat at this time from organisations such as the Irish Association of Civil Liberty (founded in 1948) and other liberal initiatives. To this threat these conservative sections reacted in various ways: positively, as in the ban on Catholics attending Trinity College, or in the campaign of 1957–58 against evil and vicious literature which was spearheaded by Archbishop McQuaid of Dublin and pursued with a will by the Catholic clergy countrywide;[6] negatively, as in the various forms of unofficial censorship, 'the whisper over the 'phone or dinner table' (Kavanagh, 1949: 130), 'the backstairs campaigns of old ladies of both sexes' (Milne, 1949: 155); or as in the different kinds of censorship listed by Seán O'Faoláin when he spoke at a meeting of the Irish Association of Civil Liberty in March 1956: censorship by fear, by bookseller, by librarians, by library users, by library committees and by the public (quoted in Adams, 1968: 150). Between 1948 and 1958, therefore, the issue of censorship was the site of a bitter struggle between those who had invested in the ideal of an organic and unchanging Irish, Gaelic, Catholic identity and those dedicated to developing a national identity which would not be exhausted by those adjectives.

The debates surrounding censorship, then, crystallised the hegemonic struggle between different strands of decolonisation, in the process helping to maintain the disabling dialectical nature of that discourse. As already noted, however, the principle of censorship, like the principle of criticism to which it is discursively related, permeates the decolonising nation in many forms, with many effects. At this point, rather than investigating examples and effects of official censorship, I wish to analyse the impact of one area of what might be termed 'unofficial', 'negative' or 'implicit' censorship by looking at some aspects of higher-educational practices in Ireland in the 1950s: curricula offered and examined, taught and researched postgraduate degrees awarded, and staff publication records in the National University of Ireland (incorporating University Colleges Cork, Dublin and Galway), Trinity College, Dublin, and Queen's University, Belfast. As the old organic and holistic intellectual spaces continued to disappear throughout the 1950s, and with the increased likelihood of the modern critic's academic affiliation, these institutions had an increasing influence on the intellectual life of the island as a whole. It is my contention that the relationships between literature, criticism and nation which were sanctioned had a crucial impact on the fortunes of decolonisation during this period.[7]

The University

The undergraduate curriculum in English-language literature at Trinity College, Dublin (TCD) was dominated by a canon of writers which, around about the turn of the twentieth century, had been hastily invoked as 'the tradition' of English literature by Oxford and Cambridge professors like Sir Arthur Quiller-Couch, Sir Walter Raleigh and Sir Henry Newbolt (Baldick, 1983; Eagleton, 1983; Mulhern, 1981). As the staple educational fare at TCD for centuries had been classics and science, the Irish university shared with those English institutions the embarrassment of handling such an uncertain 'discipline'. This embarrassment became acute with the onset of the age of mass production, mass media and consumption, for the older paradigms of value and effectivity seemed no longer applicable. What exactly was a degree in English literature supposed to demonstrate? How could one test something as subjective as a student's response to a poem or novel or play?

If a student wished for thought and contemplation there were still the classics and philosophy; if one wished for a vocational career there were the various scientific and social-scientific disciplines from which to choose. What role, then, did the study of vernacular literature play, what need did it fulfil, and what criteria – academic, moral, political – could be employed to address such issues?

These questions continue to trouble the discourse of academic literature to this day (Barnett, 1997; Ryle, 1994). For TCD in the 1950s, however, the problem of the role and function of English-language literature was compounded by the fact that the institution found itself as a small enclave representing the dwindling forces of Anglo-Irish tradition, conservatism, Unionism and Protestantism, occupying a few hundred square metres of a country struggling to come to terms with its official postcolonial status. As its weapons in this ongoing struggle were fairly limited – an occasional place on the Censorship Board, a few seats in the non-executive Senate – the university had to make use of its discrete academic faculties to maintain a voice (albeit extremely tiny) and an influence (albeit extremely marginal) in the new Ireland.

A student could study 'Irish' at TCD at this time, either as a language along with other modern European languages (including English), or as an anthropological/archaeological subject like Oriental or Scandinavian studies. 'English Literature', however, by and large meant just that – literature that was English in derivation and orientation. The 'English composition and literature' element of the 'Course for Principal and Ordinary Entrances' for TCD in 1948 was based on a selection of canonical English texts, including *Julius Caesar* by Shakespeare, various poems of Wordsworth and Keats in *The Golden Treasury (World's Classics)*, *Pride and Prejudice* by Jane Austen, *Essays of Elia* by Charles Lamb, and *Introducing Shakespeare* by G.B. Harrison (*The Dublin University Calendar 1948–49: 35ff.*). Even before she sets foot inside the institution, therefore, the would-be student is left in no doubt that 'English' possesses a political (that is, national) import above and beyond its linguistic significance. This impression, moreover, was consolidated throughout the student's first or 'Freshman' year. If studying English as part of a 'Modern Languages' degree, he was taught and examined on texts such as *As You Like It*, *Romeo and Juliet*, *The Tempest* and *Macbeth* by Shakespeare, *The Cloister and the Hearth* by Charles Reade, *Jane Eyre* by Charlotte Brontë, *Emma* by Jane Austen and *Queen Victoria* by Lytton Strachey (65ff.). If reading for a degree in 'Modern Literature' the student read a wide selection of early modern writers, including Shakespeare, Marlowe, Spencer,

Kyd, Bacon, Jonson, Browne, Milton, Dryden, Bunyan, Pope and Addison (147ff.). The individual student talent was thus left in no doubt of the 'tradition' within which he was operating whenever English literature was the subject: an organic, ongoing, conservative tradition of English letters which, owing to the peculiar circumstances of modern Irish history, was also available to certain privileged members of the Irish community unfortunately displaced from their natural cultural environment.

There were a certain number of texts by 'Irish' authors available to the Trinity Freshman, but these were limited to those 'Anglo-Irish' writers of the seventeenth and eighteenth centuries who by and large preceded modern Irish nationalism and republicanism. In fact, by invoking the works of Burke, Goldsmith, Farquhar, Sheridan and Swift against the background of a larger English tradition, the TCD curriculum hegemonically appropriated these figures into an ideal aesthetic model which formally mirrored the political and cultural relationship obtaining in Ireland during the eighteenth century. At the same time it denied the validity of any contemporary interventions which might have wished to invoke these writers as part of any alternative 'Irish' tradition. Literature in English written by Irishmen and women was valid only in so far as it could be incorporated into an older, ongoing tradition of English literature. Because it possessed no tradition, no lineage of its own, because it was a borrowed and derivative phenomenon, Irish writing in English (when it was addressed at all) must be measured against the norm of the English tradition; measured, that is to say, against those effects and attitudes invoked as 'traditional' half a century earlier by the ex-historians and ex-classicists of Oxbridge.

The nearest the TCD Languages or Literature undergraduate came to modern Irish writing in English was in the second (Junior Sophist) year when for the Michaelmas Examination she/he was required to take a 'Drama and Novel' course which included *Riders to the Sea* by Synge, *Saint Joan* by Shaw and *Juno and the Paycock* by O'Casey (69). There was, however, no context provided for these texts in the Freshman year or in the preceding Trinity or Hilary Terms; nor were they followed up in Senior Sophister year when the curriculum reverted to Restoration drama, Augustan prose and Romantic poetry (72–9). Instead, these three 'Irish' writers (all Protestant, all fully or partly ex-patriate, all at one time or another denied full national status in cultural nationalist discourse) were invoked and studied alongside the work of three mainstream English novelists: Dickens' *David Copperfield*, Thackeray's *Vanity Fair* and Trollope's *Barchester Towers*. Irish literature in English, in other words, was an interesting, somewhat exotic, backwater off the main flow of the English literary tradition, a marginal discourse recognisable by its colourful language, its improbable characterisation and its quirky depiction of the world. Having noted it, the student must make his way back to the Literary - that is, to that form of (English) writing which needs no qualifying national adjective to make it useful or valuable.

This version of 'the literary' is confirmed by the sorts of examinations set at TCD during this time. This is not only to do with the quantity and the kinds of 'Irish' writing to which students could safely be exposed (which was strictly marshalled, as we have seen). Throughout her undergraduate career the student was examined on a generic and chronological syllabus made up of texts selected from the canon of English classics. When Irish literature was invoked it was inevitably within an English frame of reference. Thus we have the following types of question: 'Write on the Irish contribution to English literature' (Entrance Scholarships and Junior and Schools Exhibitions, Trinity Term, 1952, question 6); '"The obvious later counterpart of the Pre-Raphaelite poetry is the poetry of the Celtic Revival." Mention some of the features that are common to both movements and some of the more obvious differences between them' (Examination for First Year Training College Students, Trinity Term, 1952, question 8); '"The body of native Irish literature which has been growing rapidly during the late nineteenth and twentieth centuries has made invaluable contributions to English drama." Comment' (Church of Ireland Training College Examinations Supplemental, Michaelmas Term, 1952, First Year, question 5). The student is at all times encouraged to read and understand Irish writing against the established canon of 'Literature' and all its implicit spatial and temporal discourses. The 'Irish' text thus becomes one moment in the story of English literature, and the student is invited to make sense of that text within the terms of that larger narrative and its attendant methodologies.

English Literature's supposed universalist (and thus anti-national) identity was not just a matter of text selection, however. It was also inscribed *within* both literary text and critical commentary; that is, it was a question of the 'function' of literature, what kinds of effects and values the discipline was believed to encompass, and how these might be identified and analysed by the student-critic in an examination situation. At every point in the official examination process from Entrance to Moderatorship, the student was encouraged to consider literature a primary cultural form for the expression and communication of timeless truths regarding the human condition rather than a particular organisation of writing practices subject to specific contextual factors. Hence, the following: 'On which poems would you base [Wordsworth's] lasting greatness, and for what reasons?' (Junior Sophisters Honor Examination, Hilary Term, 1952); 'Write an essay on the following: "Hamlet is a man in revolt against the sensuality, grossness and materialism of human life"' (Junior Freshman Honor Examination, Trinity Term, 1952, question 2); 'Inner conflict as a dramatic motive – its early manifestations' (Moderatorships in Modern Literature, Michaelmas Term, 1952, question 5). The effect of questions such as these is to erase history from the critical encounter with the literary text except in so far as the essentially ahistorical determinants of the human condition might be revealed. As Catherine Belsey has written in an analagous context:

> The sole inhabitant of the universe of literature is Eternal Man ... whose brooding, feeling presence precedes, determines and transcends history as it precedes and determines the truths inscribed in the English syllabus, the truths examination cadidates are required to reproduce ... When the institution of literary criticism in Britain invokes history, whether as world picture or as long-lost organic community, it is ultimately in order to suppress it, by showing that *in essence* things are as they have always been ... No history: no politics. (1988: 40, original emphasis)

In the case of TCD, this is all the more ironic because the undergraduate regime maintained at that institution mirrored the social, political and cultural organisation of which it had been deprived by history. By largely censoring at the curricular level undergraduate exposure to the range of alternative models of Irish literary discourse which had appeared over the course of the previous half century or so, the institution attempted to resist and deny the ongoing process of decolonisation on which a large part of the country had irrevocably embarked.

This impression is further confirmed if one examines the publication record of members of staff at TCD during this time. Out of 924 entries between 1948 and 1958, 397 (43 per cent) are science- or medicine-based, while 252 (26 per cent) are in the area of social science, including Politics, History, Law, Geography, Archaeology and Economics.[8] The remaining proportion is made up of 187 (21 per cent) general humanities entries (Philosophy, Religion, Classics, Languages) and 88 (10 per cent) publications in the area of literary discourse, including theory, history and criticism. Of those 88 entries, only a handful focus specifically on

Ireland and the relationship between literature and nation. Apart from A.J. Leventhal's occasional reviews for *Irish Writing* and *Envoy*, the nearest TCD gets to a sustained critical debate on the function of Irish literature in relation to the nation is an article on 'The Poetic Diction of J.M. Synge' by Donald Davie, the English poet and academic temporarily exiled at TCD and working within a specifically English frame of reference;[9] and a regular essay by the Regius Professor of Greek at TCD, W.B. Stanford, on Joyce's use of *The Odyssey*.[10] In fact, Davie's prolific output distorts the figures, as during his tenure in Dublin between 1951 and 1957 he wrote 43, or nearly 50 per cent, of the 88 texts published by members of staff at TCD between 1948 and 1958.

This counterfactual analysis demonstrates, therefore, that apart from guaranteeing the role and function of the residual Anglo-Irish section of the community, there appeared to be no forum at TCD where the issue of national identity could be analysed. By thus removing the culture/nation nexus from their own critical agenda, TCD staff were to all intents and purposes censoring debate on this issue at undergraduate and postgraduate levels and thus helping to postpone indefinitely the onset of critical practices to which such a debate might have led.

If, given the political affiliations of TCD and what it represented in the Anglo-Irish imagination, it is unsurprising to find modern Irish literature in English marginalised at that institution (at least at the undergraduate level), then it is more difficult to understand the curricular policy of University College, Dublin (UCD) during this time. The institution in the 1950s had expertise in, and courses on, Irish Language and Literature (Professor Cormac Ó Cadhlaigh), Irish Folklore (Professor James H. Delargy), Early Irish Language and Literature (Professor Rev. Francis Shaw), History of Celtic Literature (Professor Henry Gerard Murphy) and Modern Irish History (Professor R. Dudley Edwards). So, there was ostensibly plenty of scope for the undergraduate wishing to study Irish history or culture. However, as with TCD, there was no formal acknowledgement in the first two undergraduate years of the development during the nineteenth and twentieth centuries of the variety of models of the relationship between Ireland and literature in the English language. Instead, the student wishing to study anglophone literature took Pass courses on the 'Outlines of the History of English Literature', 'Outlines of the History of the English Language', 'English Prose Composition', as well as a course of prescribed reading which included Shakespeare (*Twelfth Night*), Burke (*On Conciliation with America*) and Newman (*Four Discourses on University*

Education). The Honours Student read (in addition to the Pass curriculum) some extra prescribed texts (Shakespeare's *A Midsummer Night's Dream*, selections from the poetry of Pope) and a course on the 'History of the English Language and Elements of Old English' (*University College Dublin: Calendar for the session 1948–49*: 93ff.). UCD at this time appeared to acquiesce in TCD's judgement that the place to study Irish culture was on those courses specifically designed for the purpose, for that was where the important, interesting and valuable work of understanding Irish experience was taking place. Irish literature in English was a subject for the debating society, the pub conversation, extra-curricular reading, the college magazine or the student's own imaginative work; but until the undergraduate had been given a solid grounding in those (English) writers who constituted the main thoroughfare of English-language literature, they should not be expected (or allowed) to wander off down the side-streets of Irish literature in English where all sorts of potentially dangerous complications and diversions were waiting. It was not until his final year that the UCD undergraduate had any formal introduction to Irish literature written in English when, as well as a large selection of canonical and semi-canonical English texts, the Pass student took a course on the 'Outlines of the History of Anglo-Irish Literature in the Nineteenth Century', while the Honours student was required to read all the material on the Pass curriculum plus two other courses, one of which was 'Anglo-Irish Literature 1800–1880, as represented (mainly) by the work of Moore, Darley, the De Veres, Mangan, Ferguson, Davis, MacCarthy, Fitzgerald, Allingham, Griffin, Carleton, Mitchel, Lecky' (117). In fact there appears to be a developmental or incremental principle at work in UCD's curricular policy regarding Irish anglophone literature, in which the more critically and canonically endowed the students, the more they may be relied upon to place such literature in its proper context, and thus the more of it to which they might be exposed. If, for example, the student decided to take the taught M.A. in English at UCD, she could expect a full term of lectures on 'The English Literature of Ireland' and could opt for a further term of 'Special Study – Anglo-Irish Literature of the Nineteenth Century' (171ff.); this greater exposure also made it easier for postgraduates to submit an Irish-related dissertation, if that was where their area of interest lay.

Like every other contemporary critical space, then, UCD was attempting to come to terms with the paradox of Irish literature in the English language. As much of that literature could be seen to have relevance to the contemporary political and cultural situation, perhaps it appeared judicious to wait until students had developed their intellectual faculties before exposing them to the variety of opinions on, and examples of, what should constitute an Irish literature. In the meantime, 'English Literature' signified much the same thing as it did at TCD – that is, the Oxbridge-prescribed canon. At the same time, the developmental trope noted above also appears to have been operating in the institution's own relationship with Irish literature in English, or what it continued to refer to as 'Anglo-Irish Literature'. The extra undergraduate Honours course was updated in the academic year 1954–55 to end in 1900 (106), and by 1957–58 Honours students were tested on a course entitled 'Anglo-Irish Literature 1800–1915' (113). In the same year postgraduate students taking the M.A. were given a choice between '(a) Anglo-Irish poetry and fiction in the nineteenth century, with particular attention to the poetry of Mangan and Ferguson, the *Irish Stories* of Maria Edgeworth, and Carleton's *Traits and Stories*; or (b) Anglo-Irish Literature 1904–1939: Drama – Yeats, Gregory, Synge, Colum, Murray, Robinson, O'Casey, Clarke, Johnston; Poetry – Yeats, Stephens, Colum, Clarke; Prose – Moore, Russell, Yeats, Joyce, Stephens, and the novel and short story since 1922' (146).

What this amounts to is the fact that thirty years after independence, UCD was still attempting to beat a path between liberal and radical decolonisation, unsure whether the canon of English literature it invoked was the norm to which Irish intellectual practice should aspire or whether the student's increasing exposure to Irish writing in English (and the institution's greater willingness to acknowledge that writing with new and wider courses) represented a narrative of the growing awareness of the difference between Irish and English experience. Either way, the situation was confused. This was not, however, the positive, empowering confusion characteristic of (for example) Joycean aesthetics in which the various myths of Irish history are performed so that their arbitrary (though none the less effective) natures can be observed. There is nothing here of what Richard Kearney calls 'a comic intellectual heritage typified by an ability to respond creatively to dislocation and incongruity' (1985: 11). Rather, the confusion arising out of UCD's curricular policy is part of the confusion gripping the southern state (and most immediately the state's licensed intellectuals) as it struggled to substitute an enabling *postcolonial* discourse for a disabling *post-revolutionary* one.

Another kind of hidden censorship which existed in higher education practice at this time (indeed, at any time) involved the sanctioning, supervision and writing of theses and dissertations for postgraduate degrees. As the individuals who took these higher degrees represented Ireland's educational cream and would soon be intellectually qualified to intervene in political and cultural debate, it was vitally important to be able to direct the focus of these future interventions at an early stage by controlling and delimiting the field of discourse. This kind of discursive control operated at a number of levels: the research culture of the country as a whole; institutional bias towards certain areas and subjects; individual tutorial bias towards certain models and methodologies.

Table 1. Higher degrees awarded by Irish universities between 1950 and 1958

	NUI	QUB	TCD	Yearly Totals
1950–51	62	43	31	136
1951–52	72	53	25	150
1952–53	60	47	13	120
1953–54	73	50	12	135
1954–55	31	58	12	101
1955–56	48	47	22	117
1956–57	54	45	26	125
1957–58	73	46	18	137
Total	473	389	159	1021
	46%	38%	16%	100%

Source: *ASLIB Index to Theses Accepted for Higher Degrees in the Universities of Great Britain and Ireland*, vol. 1 (1950/51)

In Table 1 we see the breakdown by institution of the higher degrees awarded in the island as a whole between 1950 and 1958. Except for 1954/55, the National University of Ireland (NUI) awarded the most higher degrees each year and the largest proportion (46 per cent) of the three institutions. This is not surprising, however, because the NUI comprised the Colleges of Cork, Dublin, Galway and Maynooth. In fact, given its greater regional availability and the larger numbers from which it could potentially draw, awarding less than half the higher degrees in the country raises questions about the Republic's educational policy, both in terms of financial resources and student performance at primary, secondary and undergraduate level.

Queen's University, Belfast (QUB), performed like a typical medium-sized British regional university with consistent returns averaging around forty-eight higher degrees awarded each year, while TCD – given its limited topographical resources and the ban in 1944 on Catholics (who represented the vast majority of the domestic student market) – started the decade well enough, fell away in 1954 and 1955, only to rally again with moderate figures towards the end of the period. Apart from these very general observations about size and resources, however, these figures do not reveal very much about the research culture in the island as a whole or institutional bias towards certain areas and subjects. However, if we break these figures down into broad subject classifications (Social Science, Science, Humanities and Literary Criticism) a more interesting picture begins to emerge.

Table 2. Broad subject classification for higher degrees awarded by Irish universities between 1950 and 1958 expressed in figures and as rounded percentages

	NUI	QUB	TCD	Total
Social Science	86 18%	51 13%	15 9%	152 15%
Science	276 58%	320 82%	102 64%	698 68%
Humanities	27 6%	6 2%	14 9%	47 5%
Literary Criticism	84 18%	12 3%	28 18%	124 12%
Total	473 100%	389 100%	159 100%	1021 100%

Source: *ASLIB Index to Theses Accepted for Higher Degrees in the Universities of Great Britain and Ireland*, vol. 1 (1950/51)

Science and technology-based degrees dominate in all three institutions and, with 68 per cent of all higher degrees awarded, in the island as a whole.[11] This indicates a country-wide move away from broadly holistic and humanistic intellectual pursuits towards a greater emphasis on specialised empirical research. Furthermore, the differing proportions (58, 82 and 64 per cent respectively) point to an appreciably greater emphasis on empiricism and vocationalism at QUB than at the two southern institutions, while the fact that TCD also outweighs NUI in this area would appear to support Liam O'Dowd's statement that 'The Catholic educational system has been biased historically towards the arts and humanities compared to the more scientific and technical orientation of its Protestant counterpart. Historically, Protestants and unionists dominated Irish science and technology' (1991: 159). The almost parity between Social Science degrees (15 per cent) and those awarded in Humanities (5 per cent) plus Literary Criticism (12 per cent, altogether 17 per cent) is an indication of the way in which scientific techniques and mores were at this time impinging on subject areas – such as History, Sociology, Politics and Psychology – which had once been the domain of the organic intellectual and the creative artist.

Overall, the figures depict a trend towards professionalism and technologisation, with the most graduates and the most research resources being attracted by science and social science subjects. For the Unionist QUB and the Protestant TCD this represents an accentuation of traditional practices, while for those postgraduates in the Republic, where intellectual activity had traditionally arrogated to itself a key interventionary role in the nation's identity, the NUI's downgrading of humanistic discourse in favour of specialised research represented nothing less than a form of intellectual censorship. This is not overstating the case, for although such a trend may have been discernible at this time in the critical institutions of other countries, these would not have possessed the same historical, cultural or political profile as postcolonial Ireland where control of such critical spaces and functions signified in a unique and highly sensitive way. Not only were there fewer opportunities to analyse the relationship between culture, criticism and the nation, but there were also fewer discursive contexts in which such analyses could be legitimately carried out. The relationship between literature and nation was retained, but was being increasingly ghettoised and marginalised beyond the point of effective intervention.

What of the higher degrees awarded in the subject areas of Literary Criticism and Literary History? The first thing to note is that, apart from reputation and charisma, individual academics had an enormous influence – deploying financial and other research resources, their own research interests, their role in the political structure of the institution and so on – on the areas, subjects and methodologies which would be admissible for research. They therefore also controlled to a large extent a student's capacity to imagine a range of alternative relationships between literature and Ireland, and the models of decolonisation implied by such relationships. Table 3 shows what proportion of successfully completed research in this area in each year contained Irish or Irish-related material.

Table 3. Proportion of Irish-related theses and dissertations in Literary Criticism/History awarded higher degrees in the universities of Ireland between 1950 and 1958

	NUI	QUB	TCD	Total
1950/51	7/13	1/1	1/5	9/19
1951/52	4/13	0/1	4/6	8/20
1952/53	10/12	1/1	4/4	15/17
1953/54	4/8	1/1	2/4	7/13
1954/55	2/6	0/3	1/1	3/10
1955/56	1/5	0/1	1/2	2/8
1956/57	6/10	1/1	0/3	7/14
1957/58	5/17	2/3	1/3	8/23
Total	39/84	6/12	14/28	59/124
	46%	50%	50%	48%

Source: *ASLIB Index to Theses Accepted for Higher Degrees in the Universities of Great Britain and Ireland*, vol. 1 (1950/51)

The 50 per cent of QUB looks quite impressive until it is seen that only twelve degrees were awarded by that institution over the entire eight years. In some years (1951/52, 1954/56) QUB had no successful submissions at all in the area of Irish-related literary discourse. These six awards mean that between 1950 and 1958 only just over 1 per cent of all successfully submitted theses and dissertations at QUB were for work carried out in the analysis of Irish language and literature. Moreover, of those six, only one – John Hewitt's M.A. Dissertation on 'Ulster Poets, 1800–1870' (1951) – began to approach the issue of modern Irish literary discourse. There appeared to be very little opportunity in the higher educational institution of Northern Ireland at this time to analyse the relationship between modern Irish cultural and national or sectional identity, an impression which contributed to the image of Unionism as a pragmatic and populist discourse which did not rely on an intellectual caste to constantly analyse and update its identity.[12]

TCD also weighs in with 50 per cent, but again the figure of fourteen Irish-related awards out of twenty-eight literary or linguistic degrees over an eight-year period is unimpressive in the light of the institution's overall research record. The highest returns were between 1951 and 1953 when eight of the eventual fourteen degrees were awarded, including material on William Allingham, Sean O'Casey, Seán Ó Neachtain, Joseph Sheridan le Fanu, Somerville and Ross, and modern Dublin theatre. In the following years, the sparse work that appeared ranged from Irish poetry in translation to Nahum Tate, J.M. Synge, James Stephens, Lord Dunsany and Lady Gregory. The overall bias of this work is *away* from contemporary developments and *towards* Anglo-Irish experience of the nineteenth and early twentieth century. The figure of 50 per cent, in fact, curiously reflects the relationship of utopian equality – half colonised, half coloniser, symbolised in the appellation 'Anglo-Irish' – which in modern Irish history is characteristic of the liberal mode of decolonisation. This relationship of equality, we recall, was imagined by the Anglo-Irish section as existing between itself and England on the one hand, and itself and Ireland on the other. So, fourteen (half) of the twenty-eight literary degrees are on Irish-related material; yet, that 'Irish' material is by and large on writers whose national identity is problematic. At the same time, the poor showing of Literary Criticism/History at the institutional level of research indicates a lack of faith in humanistic intervention and an accentuation of traditional Anglo-Irish strengths in medicine, science and law.

The 46 per cent of NUI in fact represents a much larger real number of degrees awarded for work in the area of anglophone Irish literature (thirty-nine of eighty-four) and includes material on 'The Life and Work of Francis Ledwidge' (P.B. Brady, M.A.); 'The Life and Literary Works of Daniel Corkery' (A.H. Fedel, M.A.); 'George Shiels as the Exponent of Modern Irish Comedy' (J.J. Kelly, M.A.); 'The Poetry of Jeremiah Callanan, 1795–1829' (M. Curran, M.A.); 'Charles Macklin: Actor, Dramatist, Producer' (D. Donohue, M.A.); 'An Essay on Gerald Griffin' (D. Flanagan, M.A.); 'T.C. Murray, a Critical Study of his Dramatic Work' (E.T. Conlin, Ph.D); 'Standish James O'Grady and the Irish Literary Movement' (J.J. White, M.A.); 'The Plays of Lady Gregory', (L.D. Young, Ph.D); and many more.[13] The focus of this research appears simultaneously to confirm and constitute the existence of an 'Irish tradition' stretching back to the nineteenth century and beyond.

Such a confirmation of organic continuity by literary intellectuals is typical of the radical mode of decolonisation which had come to dominate Irish cultural and political discourse around about the revolutionary period. What is new and different in the 1950s is the fact that these intellectuals had increasingly to be sanctioned by the institution, and there were thus fewer spaces in civil society where the relationship between literature and nation could be analysed. Part of Irish nationalism's traditional strength lay in its ability to mobilise intellectual discourse.[14] In the postcolonial state, however, the nationalist establishment seemed to wish to retain this ability while denying intellectual discourse the space for intervention it had enjoyed during the revolutionary period, and by means of which it had played so important a part of the anti-colonial struggle. This power, it seemed, had turned out to be a double-edged weapon as it was deployed against the nationalist establishment itself in the 1930s and 1940s by writers and critics disappointed at the apparent betrayal of revolutionary aims. Relocating intellectual activity (especially that kind of activity concerned with national identity itself) *in* the institution was an ingenious way of retaining its legitimating facility while controlling its range of influence. The NUI's overall research performance – and specifically within the area of Irish-related literary criticism – is complicit with a form of censorship in which the state attempts to control the kind and number of spaces in civil society where potentially dangerous issues of national identity can be debated; as such, it is complicit in the 1950s with the deferral of modes of discourse in which such issues might be foregrounded rather than elided or ignored.

The place where questions of national identity had, relatively recently, been raised with most force and insight – in Yeats's poetry and Joyce's fiction – is noticeably absent from all the research material of this period. Perhaps more than any other factor, this lack of institutional engagement with the work of the two seminal figures in modern Irish intellectual history indicates the regressive mode of national consciousness in which Ireland was stuck at this time.[15] Radical nationalism had performed its part in the narrative of decolonisation admirably, but now refused to yield the stage. Implicit institutional censorship of the kind examined above was a symptom of the prolonged domination of a mode of identity-formation which belonged to a past colonial era. More importantly, the ideology which informed the practices of higher education during the 1950s also played a key role in controlling the spaces where issues of national identity might be discussed, or a solution to the contemporary impasse found.

Anthologies

Another type of non-traditional critical space where the narrative of national decolonisation could be engaged at this time was the anthology. 'Anthologies', writes Francis Mulhern, 'are strategic weapons in literary politics' (1993: 23), possessing a different kind of rhetorical force from other literary texts. This force is

> the simulation of self-evidence. Here it *is* as it *was*: the very fact of re-presentation, flanked by equally self-attesting editorial learning, deters anyone so merely carping as a critic. And so, in principle, whole corpuses, genres, movements and periods can be 'finished' – resolved, secured, perfected or, as the case may be, killed off. (23)

The anthology, that is, implies a developmental, organic narrative, the ending of which is the anthological text itself and the society from which it emerges. At the same time, the anthology is a major tactic of canonical discourse, mapping the topography of seemingly already-existing literary terrains and confirming through re-presentation and editorial comment the centres and the margins of 'self-evident' traditions. What is eschewed in both cases is the *performative* nature of anthological discourse – that is, the implicit cultural narrative of the text is not presented in a finished form but is actively *re*-presented in the anthological moment; similarly, the tradition is not identified but actively constituted and *re*-created by the anthologist. The anthology, then, is not a mirror of how things *were*, but an intervention into how things *are*.

Mulhern goes on to argue that such activity has an increased significance in contexts underpinned by the concerns of national identity. The rhetoric of self-evidence which is built into anthological discourse can be harnessed to the imaginary nation so that it, too, emerges, as 'finished ... resolved, secured, perfected'. In the drive towards equality and difference, the nation produces narratives and traditions of the kind which are the staples of anthological discourse. As competition increases between different models of decolonisation, narratives and traditions which have no, or a problematic, existence in reality can be discursively recast in pristine, self-revelatory form, there to be celebrated and affirmed, offered to the nation as a reflection of how it *really* is or should be.

Modern anthological discourse developed in Ireland in the late eighteenth century under the impetus of antiquarians and collectors such as Charlotte Brooke, Charles Vallecey and Charles O'Conor the Elder. Their initiative was then given institutional form by the foundation of the Royal Irish Academy in 1785. As the nineteenth century progressed, an 'Irish cultural tradition' building on the work of earlier scholarly and historiographical intervention was consolidated, so that by the time Charles Read came to edit *The Cabinet of Irish Literature* (published posthumously in four large volumes in 1880 and again in 1906 under the editorship of Katherine Tynan Hinkson) there already existed a canon, a tradition and a narrative of 'Irish literature' on which these anthologists could draw. Reworked and re-presented by writers such as O'Grady and Yeats, this canon and the narrative it implied thoroughly charged the literary revival. The significance of this 'tradition', however, was always unstable. In works such as *A Treasury of Irish Poetry in the English Tongue* published in 1900 and edited by Stopford Brooke and T.W. Rolleston, and the comprehensive (ten volumes; 350 authors; 4,126 pages) *Irish Literature* published in 1904 and edited by Justin McCarthy, as well as in a multitude of smaller texts, the 'tradition' of Irish literature was the site of fierce hegemonic struggle as different shades of nationalist opinion looked to appropriate the canon and make it function as part of their own decolonising practice. What *was* and *was not* an Irish text? Who *was* and *was not* an Irish writer? Most importantly, what did it mean to possess an 'Irish literary tradition'? When these questions were answered 'Irishness' would be finally and incontrovertibly identified once and for all. Thus, these were the questions which set the agenda for cultural decolonisation during the revolutionary period, and which a supremely equipped anthological discourse was employed to address.

Rather than disappearing with the onset of the post-revolutionary period, however, the drive towards the identification of an authentic national essence which lies behind all these questions continued to dominate the field of Irish cultural debate into the 1940s and 1950s. On the one hand, this was because of the growth of the cult of professionalism in the humanities. When it was employed in conjunction with certain specialised scholarly practices such as editing, collecting and bibliography, anthological discourse promised, with even greater conviction than before, the discovery of an authentic Irish literary tradition. On the other hand, the genre's implicit canon-forming and tradition-confirming properties, allied with the economic, political and cultural factors examined earlier, combined to consolidate the domination of certain ways of understanding the nation, thus making it extremely difficult to put into practice (or even imagine) alternative models of decolonisation. Post-Emergency anthologies of Irish literature, then, were predominantly employed as tactical weapons to shore up radical nationalism and to render natural that discourse's highly contingent narrative of the relations between politics and culture.

Examples of the first type of anthology mentioned above were produced throughout the 1950s by scholars affiliated to, or working directly under the auspices of, the Dublin Institute for Advanced Studies. The focus of these works tended to be on medieval Irish literature, and the most advanced techniques and skills of the scholarly industry were employed to produce texts of a high professional standard. A typical example is James Carney's *Poems on the O'Reillys*, published in 1950 with 181 pages of text, 66 pages of notes and the other 68 pages taken up with a glossarial index, indexes of personal names, place and population names, poets, and first names, abbreviations, addenda and corrigenda. The anthological form itself implied a whole of which this text was a part, but with so much paratextual expertise and effort behind them, these poems on the O'Reillys appeared to 'prove' the existence of an unassailable, organic tradition of Irish literature which professional scholars like Carney would reveal more fully and with ever-increasing accuracy. In fact, the ratio of textual to 'paratextual' material, in this and similar titles, is indicative of the institutional and intellectual energy that was being dedicated at this time to the validation of the past as a crucial site for the constitution of the national identity. Such a discourse, therefore, could be incorporated into the radical nationalist drive towards pure and final essence, that utopian moment of perfect difference when the true nature of Irishness, so long 'hidden' and 'disguised', would be fully revealed in contradistinction to (although paradoxically in collusion with) the colonial other. By holding out the promise of a true Irish tradition, professional scholarly discourse in Ireland in the 1950s left itself open to annexation by the conservative and reactionary elements in the country who had already hijacked the revolution and now needed weapons to ward off the various challenges to their version of national history. With their professional aloofness and apparent conclusiveness, texts like Carney's were ideal weapons in such hegemonic encounters, as they revealed the tradition of Irish literature (and the political structures which it implied) to be a *fait accompli*.

These works, then, were 'anthologies' in the sense that they were collections of poems or fragments from the past. But they were also serious discrete works in themselves, operating at some distance from the more populist strand of the genre. Such scholarly activity, however, is always in danger of drawing attention away from the supposedly self-sufficient status of the material under consideration and towards the anthologist's arbitrary intervention. Rather than revealing the already-existing tradition, it might be said, all these notes and glossaries and indexes betrayed the anthologist's creative hand, the way in which he forged the material into a form that had no existence in reality. Such an observation, however, was unlikely to gain much credence in Ireland in the 1950s because the implied readership for such discourse was so specialised, and the value and validity of professional scholarship would have been implicitly accepted. Those disaffected and marginalised intellectuals who might have been expected to challenge the practices of scholarly tradition-forming were at this time either being recruited into the cultural institutions themselves or else had no professional experience or specialised knowledge on which to base any critique; quite literally, they did not speak the language of modern professional scholarship and thus were in no position to challenge its intellectual hegemony.

Even when the scholarly anthologies introduced a measure of self-reflection into their discourse, as in Gerard Murphy's introduction to his *Early Irish Lyrics* (1956), there was still relatively little scope outside the specialist languages and the assumed background information for someone from an alternative tradition or an alternative point of view to intervene meaningfully. Murphy's text begins: 'Irish lyric poetry is unique in the Middle Ages in freshness of spirit and perfection of form.' Now, although the rest of the text will be highly annotated, with all the techniques of modern scholarly practice brought to bear on the material to prove or argue a particular case, this first sentence is formulated with all the conclusiveness of a final sentence. The *premise*, that is, is also the *conclusion*, and rather than being the final point of the narrative it is in fact this conclusion which allows all the anthological activity – the arguments, the provisos, the nuances of scholarly discourse – that follows on from the premise. The 'is' around which the meaning turns seems to indicate a narrative that has already been told, a narrative, furthermore, which is extant, still relevant and still true. That which precedes the 'is' – 'Irish lyric poetry' – exists, is real and in the world. It is not the job of the professional scholar to question this existence, but to employ the appropriate skills and techniques to qualify and explain this existence, an operation which takes place on the other side of the 'is'. 'Freshness of spirit' and 'perfection of form' are not primary considerations at this point, as it is the existence of the tradition itself which must be confirmed by means of an assertive, aggressive linguistic formulation ('Irish lyric poetry is ...'). Murphy produces what Mulhern calls 'the simulation of self-evidence' and then goes on, by means of a highly technical discourse, to 'prove', 'demonstrate' and 'reveal' that which already exists, that which, in fact, allows such scholarly activity to have meaning and value. In this way, both modern Irish scholarship and an Irish literary tradition are secured; one guarantees the role and existence of the other by producing spaces and contexts in which each can function and have significance. The scholarly anthologies of the 1950s, in other words, by encouraging a symbiotic relationship with an ongoing, organic Irish tradition, are thoroughly implicated in the continuing domination of radical decolonisation and its insistence on continuity, essence and difference.

Such a strategy, however, is quite inefficient in the short term in that, in its original form, it is accessible only to an élite fraction of the decolonising community. To maximise its efficiency, therefore, radical decolonisation needed to saturate Irish society, reaching out into the community through different forms and more accessible narratives, thereby touching the maximum number of individuals through the most economic deployment of resources. As well as the obvious channels of primary and secondary education, another way of achieving this aim was to make use of the relatively new medium of popular mass communication: radio.

In September 1953 the state-sponsored station Radio Eireann broadcast the first of a number of half-hour lectures, the aim of which were 'to provide in popular form what is best in Irish scholarship and the sciences' (Dillon, 1959: 3). Named in honour of Thomas Davis and inspired by his famous directive, 'Educate that you may be free', many of these broadcasts were subsequently collected and published. An example of this practice is the *Irish Sagas* series, first broadcast in 1955, then collected and edited by Myles Dillon and published as an anthology in 1959 by the Stationery Office (which handled all government publications) with the aid of a grant from the School of Celtic Studies in the Dublin Institute for Advanced Studies. As well as a contribution from the editor, the text included simplified and abridged versions of various sagas from the four Cycles by, amongst others, Brian O'Cuiv, D.A. Binchy, David Greene, Gerard Murphy and James Carney – all members of the Irish scholarly élite. With the prime editorial directive to popularise in mind, in his introduction to this volume Dillon adds that 'Our purpose was to introduce listeners to the saga literature ...' (3). Having achieved its aim on an intellectual level, that is, scholarly discourse now sets out to colonise the masses and 'introduce' them to the Irish literary tradition, a tradition which has been identified, affirmed and fixed in place long before it reached this 'popular form'. Here, in accessible, reworked form are copies of originals which constitute the beginnings of the 'Irish literary tradition', a tradition which the non-specialist reader is not encouraged to follow as the further back one gets, the more tendentious becomes each stage in the 'tradition' and the less 'obvious' becomes the relationship between this cultural tradition and the political identity with which it is said to be affiliated. All that is asked of the popular reader, in fact, is that she recognise the inevitability of such a tradition and such an identity, and celebrate their existence. Whether cast in a specialist or populist register, therefore, scholarly anthological discourse in the 1950s was one of a number of useful cultural weapons with which dominant nationalism attempted to maintain the hegemony it had enjoyed in the south since Independence.

There was much anthological activity of the kind that Patrick Kavanagh would have described as 'buckleppin'' during this period. Such activity was, by and large, part of a residual liberal decolonising discourse which still looked to validate Irish experience in universalist terms. This kind of discourse could operate over a range of anthological forms, from the large 'Cabinet'-style tome to the small collection of local verse. What these texts all shared, however, was a similar model of Irish history and its relationship with the rest of the world, including the former metropolitan power. The 'unique' Irish experience which had to be produced so that it could then be incorporated into a narrative of 'universal' equality usually evoked, sooner or later, the discourse of Celticism and its depiction of the Irish as colourful, wistful, magic, sentimental – all those disabling myths invented by colonialism (and adopted by and large unproblematically by anti-colonialism) as part of its own self-fashioning discourse. At the same time, the 'other' with which Ireland must equalise came with a structured-in capacity for domination and difference, and consequently was extremely difficult to incorporate into any narrative of equality and mutual respect. In the context of the 1950s, therefore, such anthological activity was as far removed from an acknowledgement of modern Ireland's contradictory status as was radical decolonisation and its claims to organic continuity.

1000 Years of Irish Poetry was edited by the American scholar Katherine Hoagland in 1947 and ran to 885 pages. In its scope and intention it is a text which harks back to the 'Cabinets' and 'Treasuries' of the late nineteenth and early twentieth centuries. The book is divided into four parts: 'Ancient Irish Poetry' (16 authors, many anonymous, 26 translations, 90 texts); 'Modern Irish

Poetry' (20 authors, many anonymous, 19 translations, 59 texts); 'Anonymous Street Ballads' (42 texts); 'Anglo-Irish Poetry [Irish Poetry Written in English]' (155 authors, some anonymous, 378 texts). As well as this, it includes a thirty-page introduction by the editor and 'Notes on the Poets, the Translators, and the Great Books'. Such an attempt at comprehensiveness serves to consolidate the notion of an 'Irish literary tradition' – all the anthology does is make more accessible what already exists. As with most anthological discourse, the existence of the canon is not problematic, only its composition and function. Its composition, as we can see from the figures above, although rooted in medieval Gaelic poetry, is dominated by 'Irish Poetry Written in the English Language', a formulation which guarantees a role for 'English' in the Irish tradition. Its function, as ever, is to produce a realm of unique and valid Irish experience in the light of non-Irish disbelief:

> Many poets to whom I have spoken – writers who are well-known in this country and abroad, authors educated at various universities – stared blankly when told that a collection of Irish poetry covering more than a thousand years was in the process of compilation. But the long unbroken chain of Irish poetic genius has stretched over a period of two thousand years and there is documentary evidence of it in the huge manuscript volumes preserved at Dublin, Oxford, Brussels and many other renowned seats of learning. (xxiii)

The surprise of these 'writers' and 'authors' is the surprise of the non-Irish subject discovering that the subordinate subject has access to a discrete cultural tradition which, all things being equal, deserves as much respect as 'our' tradition. At the same time, this formerly hidden tradition is characterised by Hoagland in terms of a disabling Celticist discourse which, having promised equality with the one hand, snatches it away with the other:

> A mysterious unity of attitude or spiritual homogeneity identifies and marks out these Celtic people today as it did in ancient times ... A sadness pervades their genius. For all their gay songs and happy abandon, the Irish have more laments and dirges and accounts of bloody battles and defeats than have any other branch of the Celts. On the other hand, the Irish have much more humor than their brother Celts. (xxv)

Such observations derive straight from nineteenth-century Celticism, producing the same effects of spiritual elevation and political emasculation. The 'equality' offered by Matthew Arnold, for example, was no equality at all, as he attempted to flatter the Irish into a false economy of power in which they would willingly consent to English material domination so long as they were guaranteed an indispensable role in the 'more important' cultural/spiritual realm (Smyth, 1996a). And it is precisely the same species of 'equality' which is on offer here. Hoagland goes on to criticise 'the wet blanket of censorship' (liii) and other effects of contemporary radical decolonisation; but the identity founded on spiritual equality which she offers is ultimately as spurious and as disabling as the currently dominant one founded on difference. The particular Irish literary tradition identified by her, and which she sees as emerging from 'the medieval feeling for the universal character of truth', has no existence outside her own anthological discourse; but as long as it remained the only alternative to a tradition founded on discrete national experience it was implicated, as fully as the radical discourse it ostensibly opposed, in the continuing domination of colonialist attitudes in 'postcolonial' Ireland.

If such a strategy could be expected from off-shore interventions at this time, it is also easy to comprehend the strand of domestic thought which, disappointed with the achievements of the revolutionary project, fell back on myths of universalism and perfect discursive equality as an alternative to the policy of aggressive isolationism. Two examples of such alternative thought, and the different tactics they could adopt, are the Irish PEN collection, *Concord of Harps*, published in 1952, and *Modern Irish Short Stories* edited by Frank O'Connor and published in 1957. It was the former text which, despite its intention 'to afford a glimpse of Irish poetic achievement in the present century' (v), had drawn forth the scorn and anger of Patrick Kavanagh for its 'buckleppin'' propensities. What is more significant from the current perspective, however, is the contradictory double-trope which demarcates a unique 'Irish' realm of discourse and then invokes its affiliation to a realm located somewhere outside national experience. In the case of *Concord of Harps*, the first, constitutional manoeuvre is achieved by the title, which draws on cultural nationalist iconography; by the anthological form, indicating a canon and a tradition from which this selection has been made; and by the selection, dominated as it is by poets (such as Austin Clarke, Padraic Colum, John Hewitt, Mary Lavin, and so on) working within a predominantly self-conscious 'Irish' framework. The second, equalising manoeuvre is achieved (positively) by the context itself, the international PEN organisation of which it appears that Irish writers are unproblematic members; and (negatively) by the lack of any editorial reflection on the selection process, other than a comment that the poems have been chosen 'from the published work of poets who have been, or now are, members of the P.E.N. Clubs of Dublin and Belfast' (v). Because the national nomenclature was so unstable, traditional Irish anthological discourse (as found in Read, Brooke and Rolleston, Tynan, McCarthy and so on) usually included an introduction in which editors explain their selection policy. The lack of such an explanation in *Concord of Harps* underscores the impression of canonical 'self-evidence' while making no editorial concessions to a realm of 'Irish' poetry demarcated in opposition to the universal phenomenon of 'poetry'. The 'Irish tradition', it appears, is both different and equal at the same time, and this unacknowledged contradiction is precisely what characterises the discourse of liberal decolonisation.

O'Connor's collection operates in a different manner towards the same ideological effect. It is his belief, he claims in the introduction, 'that the Irish short story is a distinct art form' (xi). As he goes on to argue, however, it is distinct in exactly the same way that German folk song, the Russian and American short story, as well as the English novel, are distinct. In other words, the Irish short story is different in the same way that these other discourses are different. In his capacity as both a leading figure in post-Independence cultural discourse and a significant international writer and critic, O'Connor was familiar with both national and international cultural developments, and was thus intellectually equipped to oscillate between these constitutional and comparative narratives with relative ease. In the rest of the introduction he goes on to identify the peculiar factors – nationalism and literary history (the short story as a derivative of the novel) – that have contributed to the emergence of the Irish short story in its modern form. But once these 'distinct' factors have been identified and a discrete Irish tradition acknowledged, the discourse returns to the realm of the universal, to 'artistic and scientific truth' and the interest shared by writer and reader in fiction – 'the greatest of the modern arts' (xvii). This international writer and sometime expatriate is licensed, therefore, by virtue of his non-Irish credentials, to nominate the norm towards which a discrete Irish experience must aspire.

Both *Concord of Harps* and *Modern Irish Short Stories*, then, rely like Hoagland's *1000 Years of Irish Poetry* on a contradictory trope of difference-in-equality. This attempt to find a role in contemporary Ireland for the ideology of liberal decolonisation was unlikely to make much impact upon a nationalist tradition which had dominated the field of debate since independence and was currently in reaction against any suggestion of revisionism. Rather than broaching the possibility of an alternative relation to Irish literature, in fact, by their adherence to the traditional dialectical structure of decolonising discourse these texts were implicated in the hegemony of that dominant conservative tradition.

Another strand of anthological discourse during this time, one closely related to the last, took a broadly humanist perspective on Irish literature, combining scholarly and artistic principles to produce a literary tradition characterised by the tension between local and universal experience. Texts such as Geoffrey Taylor's posthumous *Irish Poets of the Nineteenth Century* (1951), *1000 Years of Irish Prose: Part One – The Literary Revival* (1952) edited by Vivian Mercier and David Greene, and *The Oxford Book of Irish Verse: XVIIth Century–XXth Century* (1958) edited by Donagh MacDonagh and Lennox Robinson, demonstrated no overt political affiliations, but were ostensibly interested in collecting and recording Irish literature as a unique example of a more general cultural phenomenon. These individuals were either established critics, scholars or creative artists, and it seemed that for those with the imagination and the education to realise it, the Republic of Letters was a more attractive proposition than the Republic of Ireland. The 'Irish tradition' produced in these texts is made to function in terms of more generally available cultural values and experiences – construed at various times as 'European', 'literary' or 'human'. In other words, in a register less populist than Hoagland or Irish PEN, and more self-conscious than O'Connor, Irish literary experience is raised to a higher level and made to function as a unique and valuable component of a larger cultural narrative. This refigured discursive economy is typical of liberal decolonising practice: finding the contradictions of national identity irresolvable in the world of practical postcolonial politics, literary-affiliated intellectuals invested instead in an ideal cultural world where formerly subordinate experience is reinterpreted in the light of an artificially imposed cultural standard. *We* are still different, is the message, but different in a way that *you* can recognise, in a way unique to *us* but worthy of *your* attention.

The tension between the local and the universal was reflected, first of all, in the fact that these 'Irish' authors were commissioned and published by off-shore agents: Taylor by Routledge and Kegan Paul in London, Mercier and Greene by the Devin-Adair Company of New York, MacDonagh and Robinson by Oxford University Press. Independent Irish publishing at this time was dwarfed by the achievements of London and New York, and it was still the case that success in the field of Irish letters signified little unless an individual author could win interest from a non-Irish publisher

(Adams, 1968: *passim*). Given the impressive precedents (intellectually and commercially) of Yeats, Joyce, Synge and O'Casey, and given the potentially large diasporic markets to be won, many publishing houses in Britain and America were willing to risk taking on established Irish authors.[16] Such individuals then inhabited a space somewhere between the local and the universal, employing their unique intellectual resources to translate one into the other. At the same time, it was in the interests of these individual writers to maintain the dividing line between national and non-national experience, as to a large extent their livelihoods, if not their identities, depended on the existence of these separate realms.

Mercier and Greene, for example, have no problem with identifying a peculiar 'Irish movement' – such, indeed, constitutes their text's *raison d'être*. However, they characterise this literary tradition in terms of an ambiguous binary trope in which the nation is confirmed and denied simultaneously:

> The literary movement which these forces and these writers produced is unique, if only for the fact that it achieved two things which could easily have cancelled each other. It made articulate the ideals of a people who were in the process of achieving political independence and of re-establishing their national identity, and at the same time it produced a literature capable of commanding respect independently of its geographical and political orientation ... Today students in American universities and people of culture everywhere are interested in modern Irish literature not because it either flatters or lessens the Irish in the eyes of the world, but because its achievement is formidable and significant. (1953: xx)

Rather than cancelling each other out, then, the nationalist and universalist impulses in modern Irish literature are celebrated by 'students in American universities and people of culture everywhere'. That is to say, the unique thing about the Irish tradition – the thing that makes it different – is its universal appeal, its similarity to other traditions.[17] The self-appointed role of the editors is to maintain focus on these contradictory impulses, while offering themselves as the intellectual subjects best equipped to identify one from the other. The contradictions of modern Irish identity are thus resolved in a stroke, and solved in such a way to keep the literary intellectuals employed indefinitely as they set about the now crucial cultural task of separating the unique from the general, the national from the universal.

In *Irish Poets of the Nineteenth Century* Geoffrey Taylor is not coy about the selection criteria: 'The test that I have applied to poets is that they must have been Irish by birth, and they must have written poetry with some Irish reference, either historical or topographical' (vii). This rigorous entry system appears at first to be a typical radical gesture, and Taylor is indeed initially concerned with establishing the authenticity and validity of Irish poetry in the English language. To this end, in the general introduction he invokes Thomas MacDonagh's concept of 'the Irish mode' to assert 'that there is such a thing as Anglo-Irish poetry in its own right' (vii). Taylor's anthological discourse works to confirm the existence of an organic 'Irish poetic tradition', a tradition long, large and important enough to incorporate the work of marginal nineteenth-century figures such as the ones he sets about to recover in his text. The anthology is organised into two main sections: the first is comprised of selections from the poetry of William Allingham, J.J. Callanan, Aubrey de Vere, Sir Samuel Ferguson, T.C. Irwin, J.C. Mangan and J.F. O'Donnell; the second is an 'Anthology of the best poems by less important poets'. Each poet receives a critical/biographical introduction in which their peculiarities and place in the tradition are discussed; and because Taylor, as both poet and critic, is qualified to make an intervention in this realm, the tradition is simultaneously confirmed and reinforced.

Taylor's authentication of national experience is, however, only the first part of a dialectical manoeuvre designed to guarantee local inclusion in a larger tradition, a tradition in which the political significance of *Irish* literature cannot compete with the cultural significance of the *English* language in which that literature is composed. Non-Irish experience, that is, is always implicit in Taylor's anthological discourse, always the tacit presence against which he looks to measure the emergence of 'an independent Anglo-Irish literature' (59). Thus, William Allingham's role in the Irish tradition is identified in terms of his neglect by the 'fantastically rich' English tradition of which he saw himself a part (8); Thomas Moore and Jeremiah Joseph Callanan 'were the founders of a new tradition which, since their time, has never quite lost its first flavour or been entirely absorbed in the traditions of England' (59); while Sir Samuel Ferguson 'though neglected in England and America, is certainly not negligible and has always had the ear of his own countrymen' (110). These poets, then, are representative of a discrete tradition of Irish poetry in the English language; but the very constitution of that discourse, with its dependence on contradictory Irish and English components, is always under threat, always having to be shored up by anthological invocation and editorial injunction.

Similarly, in Donagh MacDonagh's introduction to *The Oxford Book of Irish Verse*, an authentic Irish poetic tradition is evoked only to be immediately sacrificed on the altar of linguistic necessity: 'But to demand a recognizably Irish voice as a rigid test of Irish poetry would be absurd, and would exclude many fine poets. A poet speaks the language he must and that which best conveys his thought' (xxiv). As with Taylor, the only alternative to a spurious essentialist narrative appears to be an equally spurious universalist one. As we have seen, however, in a decolonising context the strategy of literary humanism was fully implicated in colonialist discourse as part of its drive towards flexible positional superiority. The unproblematic adoption of this strategy by a liberal anti-colonial practice merely guaranteed that superiority, while eroding the base on which an effective decolonising discourse might eventually operate. Taylor and MacDonagh, like Mercier and Greene, rejected the precepts of radical decolonisation, but remained trapped within the dialectical prison-house of colonialist discourse, unable to imagine a critical discourse not determined by the received categories of Irish and non-Irish experience.

Apart from *Irish Writing* and *Poetry Ireland*, which acted as kinds of regular mini-anthologies, there were relatively few collections of contemporary Irish literature at this time. *The Bell* issued a volume titled *Irish Poems of Today* in 1944, and in the same year the *Irish Times* commissioned a book of *Poems from Ireland*. The sympathetic New York publisher Garrity published *New Irish Poets* in 1948, and there was always room for one or two Irish poets, especially expatriates, in the regular collections coming from London and Oxbridge. The market for Irish literature, however, was still dominated by the giants of London and New York – Faber & Faber, Viking, Crown, Macmillan, Routledge and Kegan Paul – and the various university presses. In the limited domestic publishing culture, firms such as Hodges Figgis, the Talbot Press and At The Sign of the Three Candles were unwilling to risk too much on unproven material, and it was not until Liam Miller established the Dolmen Press in Dublin in 1951 that the work of lesser-known contemporary writers began to become more available.[18]

One notable exception to this story was the volume *Contemporary Irish Poetry* edited by Robert Greacen and Valentin Iremonger and published in London by Faber & Faber in 1949. At this time, Greacen and Iremonger were two young writers little known outside the artistic circles which they frequented. The criteria for inclusion in this anthology were that the writer must be Irish (although no definition of this term is offered) and alive, and the bias is 'towards the young and less known' (9). As with the editorial line of *Poetry Ireland* (a journal with which both had connections) the emphasis was to be less on the tradition of Irish verse and more on 'the contemporary Irish scene' (9). These younger Irish poets, that is, were more concerned with the nouns – poets and poetry – than the descriptive national adjective.

As surely as the most prejudiced affirmation of national essence, however, the apparent refusal of the younger writers to enter the debate on national identity functioned as a political intervention. The minimalist definition of Irishness constituted a denial of the realities of colonial history and the ongoing drive towards effective decolonisation precipitated by the artist-intellectuals of the previous generation. The attempt to step outside modern Irish history, or to minimise its influence by refusing to engage with it in one's critical discourse, constituted a politically retrograde strategy which, as much as any other intervention during the post-revolutionary era, prolonged the Republic's status as a culturally and psychologically colonised nation.

In any case, the 'contemporary Irish scene' was not, as it turned out, quite so divorced from the past and the Irish tradition. As Greacen and Iremonger write in their introduction: 'In order to establish continuity, however, to give an over-all picture of the contemporary Irish scene, and to indicate more clearly the principal lines of development, poems by certain writers of an older generation have been selected' (9). The 'principal lines of development' of these older poets, however, turn out to be a fair reflection of the editors' contemporary preoccupations; the main thing shared by writers such as Robert Graves, C. Day Lewis, Louis MacNeice, Ewart Milne and W.R. Rodgers is their problematic national status. When ranged alongside the work of contemporary 'revisionist' writers such as Denis Devlin, Padraic Fallon, Pearse Hutchinson, as well as Greacen and Iremonger themselves, the 'over-all picture' promised by the editors turns out to be an extremely partial image indeed. The omission of Patrick Kavanagh alone, perhaps the country's most important contemporary poetic voice, indicates the existence of an implicit critical agenda – an agenda, moreover, which represents (whatever the claims to the contrary) a political intervention.

There is, then, a fundamental tension in this text, one which reflects the tension at the heart of decolonising discourse: on the one hand, the editors seem to wish to focus attention on certain selected lines of development in recent Irish poetic discourse; on the other hand, the very selection process itself adumbrates the existence of alternative lines, while the anthological form in which they are working appears to suggest the existence of an overall organic Irish tradition. As much critical and creative discourse of the period shows, the attempt to deny Irish history was ultimately as disabling as the attempt to deny contemporary Ireland.

Anthologies remained an important form of critical intervention for Irish intellectuals after 1948. The range of political affiliation, critical approach and overall postcolonial orientation was, however, ultimately over-determined by the anthological form itself. The anthology, that is, both through the selection process and editorial reflection, implied the existence of a coherent developmental narrative and a canon of texts which could be transposed into a political register and employed to validate one or another mode of decolonisation. The scant attention paid at this time to the anthological moment itself and its performative function meant that it would have been extremely difficult to displace these canon-forming properties or the species of cultural and political debate thus allowed. These debates, instead, continued to be dominated by questions of authenticity and legitimacy, and in this way the pressing questions of postcolonialism were ignored in favour of the pseudo-issues of post-revolutionary independence.

The discourse of literary criticism in Ireland between 1948 and 1958, then, was not confined to traditional forms and spaces. Rather, a critical principle was engaged across a wide range of cultural and political spaces, three of which – censorship, higher education and anthologising – have been examined in this chapter. By thus engaging with the relationship between literature and the nation, and by offering Irish subjects various positions from which to address that relationship, these critical discourses functioned as important strategic weapons in the hegemonic struggle over which model of the decolonisation process should prevail in contemporary Ireland. The very form of these activities, however, militated against innovative thought. Especially when harnessed to the forces of reaction and conservatism, they rather contributed to the continuing domination of a specifically colonialist frame of reference for such a struggle – a frame of reference, that is, which operated in terms inappropriate to a genuine postcolonial formation. So long as Irish freedom continued to be measured in terms of difference from, and similarity to, unproblematic national identities, so long would the circle of colonial domination remain unbroken.

7 The Book

During the 1950s, some of the most influential Irish-related literary criticism appeared in the form of book-length monographs. Between 1948 and 1958, a wide range of politically affiliated intellectuals employing a number of different critical approaches published volumes of literary history, collected essays, reworked theses, genre analyses, critical surveys and biographies. As with periodical criticism and the various non-traditional critical forms, such activity constituted highly motivated interventions into the debate surrounding the relationship between literature and nation, and thus the nature of postcolonial Irish identity.

The single most influential factor on the development of book-length Irish-related literary criticism during this period was the growth of international academic and general intellectual interest in modern Irish literature. While there had been a certain amount of curiosity about this phenomenon – especially Yeats and Joyce – since the 1920s, it was in the postwar period that this curiosity began to take institutional form, paving the way for the development of a full-blown critical industry involving academics, universities, publishers, book-sellers, tourism policies and advertising agencies.[1] If, as was widely acknowledged, modern Irish literature constituted one of the most important international cultural developments of the twentieth century, then despite the geographically specific nature of much of it, the fact that this literature was written in English made it fully available to critics in Britain and America. The enormous output of Joyce, Yeats, Synge, O'Casey, Shaw and so on, constituted potentially fruitful intellectual territory – territory, moreover, that apart from some minor local incursions was practically virgin and ripe for scholarly colonisation.

If one was looking for the moment of take-off for the international invasion of Irish literature, one could do worse than settle on the publication in 1948 of Richard Ellmann's book *Yeats: The Man and the Mask*. Ellmann went on to publish another book on Yeats in 1954, and throughout the decade was working on his massive critical biography of Joyce. Along with critics such as Robert Deming, Norman Jeffares, Hugh Kenner, Frank Kermode, David Krause, Richard Kain, William Tindall, F.A.C. Wilson and many more, Ellmann represented the international scholarly élite which began to dominate the study of modern Irish literature at this time.[2] These developments had a number of important implications for Irish criticism and the ongoing process of decolonisation with which it was directly linked.

'Industry', albeit a cultural industry, signifies competition, and competition signifies an inevitable hierarchy of critical practitioners, subjects and methodologies. In the heightened activity of postwar Irish-related literary criticism, certain ways of thinking and writing about Irish literature grew dominant, while others became marginal. The first to feel the competitive pinch was the small body of Irish intellectuals who were finding it increasingly difficult to offer their own national experience as a legitimate critical criterion. The scholar-critic of the typical Plains University of America did not need to have walked the streets of Dublin to produce an interpretation of *Ulysses*, nor to have surveyed the landscape from Ben Bulben to analyse 'Sailing to Byzantium'. The 'New Criticism' which constituted the dominant Western theoretical perspective on literature in the years after the Second World War emphasised the self-sufficiency of the text, and eschewed interpretations which relied on historical, biographical or contextual detail. The typical British or American academic critic, therefore, was unlikely to stress local experience in their subjects' work, or to invoke historical knowledge in their own, and thus less likely to adapt nationally specific perspectives in either. As the international domination of Irish-related criticism grew, certain strategies became increasingly marginalised; even over the short period in question here, it is possible to see the emergence of different kinds of critics of Irish literature, performing different kinds of tasks for different kinds of readers. Nationality was no longer an automatic passport to critical attention, nor to critical marketability, and with the fairly restricted Irish publishing industry to consider, the average Irish critic of the 1950s faced an intellectual and material dilemma.

This brings us to the second major implication of this international literary industry – the impact on the range and type of critical approach encouraged by such developments, and the debates surrounding the relationship between literature and the nation. In its traditional form, the book-length treatise lacked the immediacy of the journal, and was less insidious (and thus, arguably, less effective) than the various covert critical interventions, such as censorship, pedagogy or anthologising. At the same time, the extended encounter between critic and reader allowed the former to produce prolonged, reasoned arguments at various levels of sophistication, and thus to engage the reader much more fully. The balance between thesis and example could be extended to cover a much greater range of material than that available to the journal article or the editorial introduction. This allowed the critic to build up a model which, with every example and every development in the argument, invited the reader to acquiesce with the critic's own understanding of the function of literature and its relationship with the nation. The book, that is, offers the reader a full and coherent realm of discourse – a realm which, operating stylistically, tonally and conceptually, could be transposed into a practical narrative of national development. The reader was invited to discover the 'truth' of the book's argument for herself through an extended encounter with a range of discursive tactics. The critic, that is, could exploit the book's key formal property – its length. By allowing the book-reader to become active, to anticipate points, to make connections between arguments, to follow threads and nuances over an extended period, to appreciate the organic growth of a position, the book-writer could utilise a factor not available in the shorter forms.[3]

When cross-fertilised with the growing international element in the discourse, these formal strategies made for a critical practice which tended on the one hand towards hagiography and on the other towards esoteric theoretical exposition. Criticism in the postwar world was becoming a full-time intellectual occupation. It was no longer a hobby for the creative writer between novels or poetry volumes, but an undertaking demanding professional attitudes and qualifications. Firstly, scholarly editions of the Irish writers had to be produced – a task which, given Joyce's publication history and the creative practices of other canonical figures, was by itself a full-time career. O'Casey's collected plays began to appear in 1949; *The Stories of Frank O'Connor* was published in 1952, and *The Stories of Liam O'Flaherty* in 1956; collected editions of the work of Oscar Wilde appeared in 1948 and of James Stephens in 1954, while the work of Swift and Yeats appeared in many volumes throughout the decade; scholarly editions of Shaw and Synge began to appear just at the end of our period. These works had then to be interpreted, analysed, explicated and evaluated in ever finer detail, to ever greater extents. In relation to the changing critical practice surrounding Joyce at this time, Bruce Arnold writes of

> the shift in the character of the critical attention and analysis away from the broadly-based considerations of style, character, plot, narrative skill, humour, irony and satire, which had generally been the collective headings under which his writer critics assessed him, into a rather

> different academic approach, which increasingly considered myth, symbol, allusion, pun, word-play and textual analysis. (1991: 95)

While discussing the critics who engaged in such discourse he writes of

> a quite distinct though associated shift from the broad mixture of favourable and unfavourable judgements ... to an increasingly adulatory critical starting-point adopted by the predominantly academic people who had made an often irreversible lifetime commitment to Joyce studies. Such people ... start out from a position which largely removes from their critical canon the possibility of finding against Joyce the writer in any serious or fundamental way. (95)

The same is also largely true of Yeats studies, and the other schools surrounding the 'minor' figures of modern Irish literature. New, specialised critical tools were being brought to bear on these texts, tools which demarcated the range and nature of engagements possible with an Irish literary text. Subjective responses which included anecdote, appreciation or evaluation based on restricted local experience were marginalised by a critical discourse concerned with general principles of interpretation and universal experience. The newly hegemonic critics neither possessed nor desired any sentimental connection with a national audience. So whereas Yeats may have deliberately intervened in questions of national identity from the stage of the Abbey, this latest generation of critics disdained this capacity, writing instead for a limited audience while rigorously vetting would-be interventions in the light of the currently dominant critical criteria. At the same time, as Arnold notes, this new critical practice came with a built-in evaluative property, where even minor pejorative formulations have to be contained within an overall celebratory framework. It thus becomes illegitimate to condemn an Irish writer on political grounds, if the critical practice in which such a judgement was cast depended on that writer's continuing currency. A typical example is F.A.C. Wilson's *W.B. Yeats and Tradition* published in 1958, in which he writes:

> These essays ... are interpretative; and my attempts at evaluation are no more than casual. I have usually given my own estimate, as for example of *The Herne's Egg*, but I have done so in full awareness that I may have overpraised the work. A result of the New Criticism, or so it seems to me, is that Yeats's stature as a poet has been too rashly guessed at; his poetry has been evaluated before it has been perfectly understood. The present essays may be a first step in the reversal of this process: I have tried, at all events, to give a definitive statement of meaning, in

> the hope that another critic, who will be concerned with evaluation, may use my work as a foundation upon which to build. (9–10)

The critic cannot evoke national criteria, as this would disbar his own intervention on grounds of legitimacy; neither can he attack his subject, as this would curtail the scope of future engagement. The best he can do, as Wilson does in this passage, is eschew, apparently, evaluative criteria while focusing on those specialised techniques which demarcate his area of legitimate intervention. Although written from a 'frankly sympathetic' (9) position, this scholar-critic looks to reinterpret Yeats's work in the light of new, specialised research (the Platonic tradition of subjective philosophy) while maintaining the aura of an aloof professional concerned only with correct interpretation. The apparent denial of critical judgement is in fact disingenuous, a subtle tactic to disguise by anticipation the generally positive evaluation which informs the whole discourse, from choice of subject to the apparently value-free 'definitive statement(s) of meaning'. If local in inspiration, Yeats's work is universal in import, and any critical intervention which fails to recognise this 'fact' will be wrong. The intended reader is familiar with New Criticism, perhaps the other critic 'who will be concerned with evaluation' once the important foundational work has been completed; in this way, Wilson locates himself among a network of similarly concerned critical subjects, operating in a sphere far removed from questions of national identity or cultural decolonisation. At every point, therefore, the critical moment intervenes in the literary moment, determining what kind of writing should be considered and how it should be judged.

Such a critical practice, therefore, had important ramifications for Irish decolonisation during the 1950s. The monograph itself, with its greater length and capacity to engage a reader's intellectual faculties, could be employed as a tactic in the struggle between the various versions of national identity. At the same time, the less immediate impact of the book compared to the journal article or the more popularly affective forms of critical intervention, meant that its range of intervention was limited. As a consequence, the dialectical model of decolonisation which had dominated modern Irish-related literary criticism was unlikely to be questioned. Moreover, when this critical book-writing began to be dominated by an international scholarly élite, any Irish intellectual who failed to adhere to the agenda of professional academic criticism on the one hand, and generally favourable evaluation on the other, risked marginalisation. The connection – at once sentimental and problematical – between critic, writer and audience so prized by the intellectuals of the revolutionary period was sacrificed to universalist principles which allowed the off-shore subject a dominant role in Irish critical discourse.

Irish critics working during the 1950s reacted to these developments in a number of ways. Whereas critics in the immediate post-revolutionary years had found it difficult to escape the dialectical impulse of decolonisation with its twin narratives of similitude and difference, the dominant trope in the new criticism was universal rather than local. The debates surrounding Irish identity were therefore secondary in importance to questions of general human experience as mediated by literature. On the one hand, by removing attention from the culture/nation nexus, the new criticism facilitated the maintenance of the old order, leaving liberal and radical modes intact and still dominating the field of legitimate intervention. On the other hand, this same manoeuvre could work to displace that very field of debate, so that universalist interpretations delivered by non-national critics might help to problematise the very notion of an Irish tradition and the national identity which it supported. Much depended on the intellectual make-up of the particular Irish critic and the contexts in which she was looking to intervene. In the rest of this chapter I shall examine a number of such interventions and consider the implications of certain general developments in book-length Irish-related criticism for political and cultural decolonisation.

The 'Life and Culture in Ireland' Series

In February 1948 southern Ireland's first coalition government took office, comprising representatives from Fine Gael, Labour, Clann na Poblachta and a few Independents. One of the most dynamic figures in this new administration (the first change of leadership since 1932) was Séan MacBride, founder and leader of the radical republican party, Clann na Poblachta. Opinion on his achievement and impact on Irish life in his relatively short political career is divided; however, during his tenure as Minister for External Affairs he was responsible for establishing one of the most interesting experiments in postcolonial government when he appointed and launched the Cultural Relations Committee of Ireland in January 1949.

The Cultural Relations Committee was established 'to consider schemes with special reference to their practical value in relation to the public expenditure involved, for the promotion and development of cultural relations between Ireland and other countries, and to advise the Minister of External Affairs in relation thereto'.[4] The Committee's activities included film commissioning, photography exhibitions, student exchanges, musical tours and so on. Like any other institution with a budget in Ireland in the 1950s, the deployment of its resources was a constant bone of contention.

One of the main developments by the Committee was to commission a series of monographs by some of the country's foremost cultural figures to give, in the words of the introductory note which prefaced each volume, 'a broad, vivid and informed survey of Irish life and culture, past and present'.[5] The purpose of the series was to commission suitably qualified individuals from each field to cover a wide a range of cultural material for the largest possible audience. Despite the disclaimer which came with each volume – 'Each writer is left free to deal with his subject in his own way and the views expressed in any booklet are not necessarily those of the Committee' – these titles represent a sustained attempt by the state to enlist a number of non-state-affiliated intellectuals in the specific hegemonic task of justifying and naturalising a particular view of Irish history and culture. 'Culture', in fact, was one of the few growth areas of modern Irish life. In the face of growing international interest in recent Irish cultural experience, the state could not (literally or figuratively) afford to let it circulate unmarshalled. So, while the Cultural Relations Committee series may have seemed like a willingness on the part of those in power to review national identity as shaped by cultural experience, it actually represented a belated effort on the part of the state to produce legitimate credentials in an important area of national experience, while simultaneously looking to intervene in the range and nature of that experience. As such, it is a typical strategy of the nationalist bourgeoisie, indicative of the need to harness important decolonising discourses such as literature and remove their potentially dangerous capacity to intervene in matters of national importance, while retaining their prestige and affective powers.[6]

The individuals who participated in this undertaking came from a number of intellectual disciplines and possessed a range of sectional affiliations. The dearth of outlets for cultural discourse in Ireland at this time has already been remarked, and the chance to publish a volume for the Cultural Relations Committee was no doubt a welcome offer for most of these writers. The dilemma then facing each contributor was whether the putative freedom promised at the beginning of each volume was enough to balance the loss of intellectual freedom which came with being recruited by the state to perform an important political task. Different contributors confronted this issue in different ways: some looked on it as an opportunity to circulate their work in a more popular form; some took the opportunity provided by the disclaimer to attack contemporary political attitudes and practices; some were quite happy to be so affiliated. Each strategy represented part of the struggle over the function of culture in relation to nation in contemporary Ireland, and these struggles were played out not only in the argument of each volume, but on the crucial levels of form, tone and style, critical approach and targeted audience.

The books were usually quite short and intended to be neither populist nor specialist. They were written for the most part in what would be known in modern parlance as a 'user-friendly' style. In the texts which deal with culture in a written form (numbers I, II, VI, IX and X) there is no original research but rather statements of opinion or restatements of established knowledge in more accessible language. The words 'broad' and 'informed' seemed to be aimed at the intelligent layperson. However, this left an enormous amount of scope for individual interpretations of that imagined subject's capacities.

Tending towards the 'informed' end of the register, it would be extremely difficult to write about some topics, such as medieval poetry, in simple terms without undermining the often nuanced arguments and specialised critical approaches with which the discourse functioned. Even in the title of her contribution – *Irish Classical Poetry, Commonly Called Bardic Poetry* (1957) – Eleanor Knott reflected the contradictory nature of Irish culture as conceived by modern scholarly discourse. In the terms of that discourse, Ireland did indeed possess a classical poetic tradition, currently in the process of being reclaimed from the populist narrative which had come to dominance in the last years of the nineteenth century. An authentic tradition, linked to an authentic identity, underpins contemporary specialised research; yet, in spite of the 'general reader' invoked in the prefatory note, the discourse disdains contemporary sanction by the inheritors of that tradition and that identity. Significance still turns on highly esoteric distinctions – style, tone, meter, rhyme, genre, grammar, allusion and a great wealth of secondary scholarship. An unbridgeable gap which is 'identified' at a conceptual level (between classical and popular poetic traditions) is confirmed and reconstituted by the contemporary critical metadiscourse. Furthermore, in spite of the highly technical language she employs, Knott's discourse crucially turns on a basic and easily apprehensible distinction – one unique to Ireland – between poetic ability and political disability. For example, responding to the interpretation of the legendary 'Contention of the Bards' as a political event by the nineteenth-century scholar Eugene O'Curry, she writes: 'In any case it remains for us a treasury of idiom, of poetic style, of legendary history and tradition, however pathetic as a quasi-political enterprise it may now appear' (78). Like Matthew Arnold (whom she cites on page 18), a sense of style and dignity are offered as the distinguishing characteristics of Irish poetic discourse; and this formulation disallows the tradition a political dimension, either at the primary (creative) or the secondary (critical) level.

On the other 'broad' hand, some arguments specifically relied on their accessibility and simplicity, on the straightforward common sense which appeared to inform them at a conceptual and formal level. In *The Fortunes of the Irish Language* (1954) Daniel Corkery rehearsed the arguments which he had introduced some thirty years earlier in his books *The Hidden Ireland* (1924) and *Synge and Anglo-Irish Literature* (1931):

> The tradition of the Irish people is to be understood and experienced with intimacy only in the Irish language. It would be impossible that it could be so come upon in the English language ... To say tradition is to say language – and while this is true of every national tradition it is overwhelmingly true of ours ... Soil: Literature – where else, in Europe, do these words go almost inseparably hand in hand as they do in our tradition right down the centuries? (13–14)

By postulating an organic link between culture and politics, between poetry and the people, Corkery's discourse works in a different way from that of Knott, although substantially towards the same end: towards, that is, the confirmation of a continuous cultural tradition founded on a distinct, unique Irish identity. Whereas Knott had posited a gap between high and low culture, and had replicated this gap at a formal level in her own discourse, Corkery demonstrates the lived continuity of experience in Irish culture and politics by pitching his own argument at a straightforward, non-technical level aimed at the ordinary Irish person. In Roland Barthes' terms (1972: 650), Corkery 'covers' the primary, 'real' relationship between Irish language and life with his own secondary, 'unreal' discourse. At the same time, the fact that he must argue these truths in a written form is an argument against natural convergence between language and life in Ireland, past and present. This relationship between a 'good' reality and a 'bad' representation of reality recalls Derrida's critique of Plato's contradictory strategy in the *Phaedrus*, where the ancient philosopher's attack on the secondary nature of writing as opposed to primary reality was constantly undermined by the fact that this very argument could appear only in the form of a written discourse (Derrida, 1981: 61–171). The problem, for Plato and Corkery, becomes one of audience and style: *for whom am I writing, and how can I represent naturally occurring phenomena through a secondary, supplementary discourse (that is, through technique, narrative, argument and so on)?* After the initial point is made, there is very little left for Corkery to do except re-formulate and re-present the organic nexus of Irish life, language and literature in ever clearer, ever more self-revelatory, terms.

In both cases, then, the critical tactics employed were sympathetically related to the effects desired. The overall effect of this relationship was that no matter the deconstructive impulse behind an individual intervention, congruity of form and content militated against any critical model in which these categories might be considered from an ironic or discordant perspective. Both the conception of the series and the execution of the individual volumes, therefore, were unlikely to amount to much of a challenge to the traditional model of decolonisation and its limited choice of national identity.

This impression is confirmed by one of the major titles in the series, Austin Clarke's *Poetry in Modern Ireland* (1951). Clarke was one of the dominant literary-affiliated intellectuals in the post-revolutionary era, and had for some years been considered by many to be the major poetic voice in Ireland since Yeats. Although not difficult or technical, his critical discourse is wide-ranging and knowledgeable, implying a readership already familiar with certain developments in Irish cultural and political history. At the same time, it is difficult to discover any informing argument or position in the text. Clarke traces a narrative of poetic development in modern Ireland, beginning with the Yeats-inspired Celtic Twilight, the use of myth and folk art, the Gaelic revival and the various challenges to, and experiments with, the idea of an Irish tradition. Throughout this narrative, the author appears to be playing the 'grand old man' of Irish letters (although he was only fifty-five at the time of writing), relating a narrative from which he, as organiser and orchestrator, remains coolly aloof.

The book is one of the shorter titles in the series (seventy-one pages), especially in comparison with some of the volumes which were being produced in the area of Irish-related literary criticism in Britain and America at this time. Clarke was primarily an artist rather than a critic, and much of the volume is taken up with quotation and example. What remains is very much a set of personal impressions, almost diametrically opposed in intent and effect to the highly scholarly, well-argued tomes of the academic critics. At times the text is no more than a list of names linked through critical nomination with large abstract concepts such as modernism, myth and nation. It is not amenable to extraction, as there is hardly any passage in which the author sustains a critical point or risks any kind of judgement, subjective or scholarly. At times, in fact, the text reads more like a catalogue than a work of literary history. This aloof personal style, however, does not represent an attack on current practices; rather it is an attempt to translate artistic detachment into critical discourse. By apparently refusing to invest subjectively in Irish cultural debate, Clarke the poet-critic attempts to step outside history. In doing so, however, he fails to appreciate that such a manoeuvre can only take place *in* history and that in spite of its bland tone and apparently non-committed stance, *Poetry in Modern Ireland* in fact constituted a highly political intervention into the debate on decolonisation with which its subject matter – Irish literary history – was already linked. It is as if the poet-critic has seen through the myths of liberal and radical decolonisation and withdrawn into an unassailable, because uncommitted, position where his only affiliations are to himself and his art. Clarke attempts to locate a position outside Irish literary history, for only in this way can he remain uncontaminated by the spurious identities offered by that history. However, whereas Joyce's exile had been physical but not intellectual, allowing him to attack dominant decolonising practices while retaining a stake in Irish history, Clarke's critical exile works towards the exact reverse. His short-term tactic to salvage a workable position for himself as poet and critic could be absorbed by a long-term strategy which confirmed colonial similarity and difference as the only alternatives on which to base Irish tradition and identity. The fact that, as one of Ireland's more famous modern writers, he has been unproblematically recruited by the state, only compounds this interpretation.

Clarke's lacklustre performance was all the more disappointing because in terms of the challenge to dominant decolonising practice and its limitation of the debate to variations on sameness and difference, the series had started quite auspiciously. Micheál MacLiammóir's *Theatre in Ireland* (1950) is a sustained, reasoned and witty attack on nationalist isolationism at both material and intellectual levels. As co-founder with Hilton Edwards of the Gate Theatre in Dublin in 1928, MacLiammóir was predisposed to reject the isolationist line of post-revolutionary cultural politics. To balance the dominance of the Abbey and its representation of Irish life solely in terms of cottage or tenement, throughout his career MacLiammóir fostered an alternative practice which would offer 'a totally different conception of the theatre and its relationship to Irish life' (25). He took the opportunity offered by the Cultural Relations Series to write a history of Irish theatre which would function at the same time as a critique of contemporary attitudes and practices.

MacLiammóir's book is in fact a classic case of the postcolonial intellectual who can identify the dangers of sustained national investment in radical decolonisation but is ultimately unable to imagine a realm beyond the parameters of colonialist discourse. Writing of Yeatsian nationalism, he says:

> But when, in later years, the nation having (most unexpectedly) obeyed the Yeatsian call, having indeed taken it so literally that for some time it seemed that she would never again be able to free herself from the engrossing contemplation of her own perfections and peculiarities, then came the time to open the windows once again and look, not at the glimpses of a world that England chose to show, but at the world itself. (21)

Here, MacLiammóir apparently rejects the binary restrictions of colonialism in favour of the larger experience available from contemplation of the political and cultural landscape beyond traditional experience. As astutely as Kavanagh or O'Faoláin, he recognises the fact that contemporary Irish attitudes are trapped within an outdated system of thought; like these critics, however, he finds it extremely difficult to escape that system himself, and his discourse is constantly being drawn back into a dialectical register, as in the opposition between 'the nation' and 'the world' in the above extract. At times, this dialectic is invoked in Kavanaghesque terms, when he posits a self-conscious, ironic relationship between native and universal sensibilities, as in the following statement: 'It is in the expression of profoundly local and intimately known types and places that the dramatist comes closest to the secret of universal portraiture' (38); or when he invokes an organic link between Irish art, criticism and life, as in the following:

> The audience must be led towards a fuller understanding of the art of the stage, an art which may yet be doomed to eclipse by that of the screen unless we discover a new and varying series of forms and expressions, and here a great school of criticism might serve as a guide. (45)

Note the characteristic turn of the alienated postcolonial intellectual to criticism, to the *constitutional* powers of secondary interpretation to provide a space for the primary *creative* discourse which will signal the opening of Ireland's windows on to the world.

As with Kavanagh, O'Faoláin, Knott and Corkery, MacLiammóir's tone (urbane, informed, restrained) is in complete accord with the position (liberal, open-minded, cultured) he is defending. What carries him further along the road away from a bland universalist alternative to the identitarian politics of traditional decolonisation is his wit and a self-consciously 'camp' style:

> Ireland, who after many centuries of preparation for a national rebaptism has groped her way to the church door and is at least in sight of the font, is also, unless she exerts every muscle, in danger of falling into the dazed reaction of a prisoner who has waited too long for his release, into the complacency that will produce nothing but eggs and butter and bacon for a world that may rightly expect far more. (42)

Here, metaphors are mixed, images are evoked and language is employed in a explicitly 'literary' style that exceeds its communicative, critical function. 'Primary' or creative discourse is invading 'secondary' or critical discourse, threatening to displace that particular opposition and its multitudinous political and cultural hypostases. Given the degree to which his life was shaped by English background and concerns, there is a certain element of 'passing' (with all the ambivalent identity politics which that brings) animating all MacLiammóir's 'Irish' work.[7] At the same time, with his practical experience in the theatre and his appreciation of the potential of role-playing to disrupt normal, 'real' identity, perhaps MacLiammóir was in a better position than his literary-affiliated fellows to launch a critique against contemporary Ireland. The critique is not sustained and often lapses back into a universalism which merely reverses, and thus reprises, dominant decolonising practices. Like Kavanagh, however, MacLiammóir's discourse affords glimpses of the difficult and dangerous freedom imagined by contemporary postcolonial theory. Written from within a state-affiliated initiative, *Theatre in Ireland* at least demonstrates a partial grasp of the need to seize the rules of dominant discourse and rework them so that they can be redeployed for alternative cultural and political ends.

The 'Life and Culture in Ireland' series initiated by the Cultural Relations Committee was an interesting development in Irish cultural debate in the 1950s. As populist alternatives to the scholarly discourse of international criticism, these volumes represented the state's wish to maintain some degree of influence in a crucial area of national experience. MacLiammóir began the series by showing how the political ideology behind this wish could be exposed and perhaps turned back against the state and its compromised model of Irish identity. Even his challenge, however, was limited and curtailed by the form of his intervention and the context in which it was undertaken. The contributors following him appeared to have less understanding of the problems involved, and/or less commitment to solving them.

Truth and Method

As scholarly and professional as it might be, modern international criticism could not compete with the standards maintained by that strand of domestic critical practice interested in Ireland's medieval and early modern culture. As noted above, monographs like William York Tindall's *James Joyce: His Way of Interpreting the World* (1950) or Frank Kermode's Yeats-based *Romantic Image* (1952), although scholarly texts written for a restricted audience, could work either to support or to problematise the dominant decolonising system of thought. This was because the universalisms usually propounded in these texts were often counteracted by their authors' foreignness – thus fracturing the usually clearly demarcated borders between Irish and non-Irish experience.

The books which appeared throughout the 1950s under the auspices of the Dublin Institute for Advanced Studies or were published independently by individual members of the Irish scholarly élite presented no such ambiguity. The discipline represented by this body, to recall, was comprised of literary, linguistic, historical, archaeological and anthropological elements practising a highly specialised form of scholarly research. Moreover, this discipline had attracted international interest much earlier than its modern counterpart, with scholars such as Renan, Meyer and Thurneysen apparently giving the lie to the nationality/authenticity nexus.

Whatever the national origins of the individual scholar, however, the technical languages, the indices, the references, the acknowledgements, the finely nuanced arguments, the appendices, the bibliographies – all this activity took place for one reason and was geared towards one goal: discovering the *truth* of the Irish tradition. No matter the debates over Yeats's use of myth and mysticism, no matter Joyce's experiments with language and form, no matter these and all the other questions surrounding *modern* Irish literature; the critical discourse brought to bear on these issues was ultimately just a matter of opinion, for there was no final court where these questions could be settled once and for all. However, while individual practitioners might have acknowledged that their discipline operated in much the same way, the discourse of historical scholarship depended on the never-relenting drive towards ever more accurate and ever clearer interpretations. It depended, that is to say, upon the concept of a final and true version of Irish history waiting to be discovered and brought to the world by the most talented and conscientious scholar. The promise (implicit and seldom formulated) was that this historical truth would be reconnected to the present at some point, and the specialised scholarship was therefore just as relevant to contemporary Ireland as the more accessible critical discourses. In the meantime, the (re)search continued.

A book like James Carney's *Studies in Irish Literature in History* (1955) is a good example of the sorts of issues surrounding book-length historical scholarship. Before the 'argument' or subject of the book can be confronted, the reader must negotiate a number of paratextual boundary-markers which signal this text as synecdoche for a wider undertaking, one that simultaneously informs and allows single interventions such as the parallelepiped in the reader's hands. Firstly, the contents list comprises titles and subtitles the significance of which would be lost on the general or uninitiated reader – for example, Chapter VII, which reads: 'The Féuch Féin Controversy (243); Féuch féin an obair-si, a Aodh (243); Gabh mo shuirghe, a ua Émuinn (266)'. Obviously, the reader is expected to bring some extra-textual resources to the text (in this case, critical and linguistic) if he wishes to engage fully with the argument. That same reader is then confronted with two full pages listing 'Abbreviated References' (x–xi), citing forty-eight primary and secondary sources in Latin, Irish, English, German and French, such sources to be found in the libraries of Ireland, England, Europe and America. This text, then, is but a single enunciation in a large ongoing conversation, and if the reader is not aware of the codes and assumptions on which that conversation is based, she is discursively disabled, unlikely to catch the significance of this particular intervention. Finally, the reader notes that in the 'Foreword' certain prestigiously titled individuals – Dr Michael O'Brien, Professor Eleanor Knott, Professor Myles Dillon – are acknowledged for their 'valuable suggestions ... advice and criticism'. The author has felt it important to thank these individuals, but not others. But is this important? Is their work important? And how does their work relate to the text the reader holds in his hands?

In each of these ways the reader is being positioned, depending on their response to these boundary-markers, on either side of a divide which separates legitimate and illegitimate interventions in the debate surrounding 'the Irish tradition'. And the reason behind all this activity, all this discursive demarcation and policing of personnel, is accuracy, greater probability, heightened certainty: that is, the thing for which all these are merely surrogates – truth. Of course, the 'truth' might be that there was no such thing as an 'Irish tradition', cultural or political. In the meantime, however, this search for truth, carried on at such a sophisticated level, gave intellectual sanction to the concept of a final, finished Irish history based on a final, finished Irish identity. This all-informing concept in turn supported the system of thought which insisted that the only legitimate choice in the matter of decolonisation was between similarity to, or difference from, the former colonial power.

The drive for truth lent the discipline of historical scholarship an almost priestly aura. As part of a uniquely privileged institution within the state, the political affiliations of the individual scholar were firstly with the institution itself and secondly with the state which partly funded, and wholly sanctioned, that institution. The state granted the scholars a valued social space to carry out its arcane mysteries and the status that went with professionalism. However, as most of these individuals were academics or employed by the scholarly institution in some capacity, there was no official self-critical strand within the discipline. To intervene in the discourse one must learn its languages and procedures, its codes and practices. Any critique offered in these terms, however, could be absorbed back into the discourse, and the intended challenge would end up actually reinforcing the boundaries of the discipline as well as the social and intellectual power it wielded. There was, in other words, no recognised space outside the discourse where its political affiliations could be inspected, or its rationale analysed. The discipline could generate controversy within its own ranks, but these were only local and generational disputes; the overall search for truth and the institutional location of that search remained fundamentally unchanged. Rather, the period between 1948 and 1958 saw a shift in Irish intellectual hegemony, away from the type of holistic intervention coveted by contemporary writer-critics, and towards state- and institutionally-affiliated scholarship.

This retreat into an official search for truth was partly a response to the domination of the field of Irish-related critical discourse by an élite band of international, career-minded, academic critics. Whereas Joyce-studies or Yeats-studies required whole new critical languages – languages which the American and British universities were inventing or well-equipped to learn – the techniques and procedures of historical scholarship had remained largely unchanged since the nineteenth century, and the progression from undergraduate to postgraduate to researcher was easy to follow for the Irish scholar. The growth of book-length historical scholarship in the 1950s, in fact, might be seen as a response to rising international interest as it can be argued that it was under the pressure of a large, lucrative critical industry churning out monographs on Irish-related subjects that Irish historical scholarship turned to a form – the book – for which it was not really suited. The natural form for historical scholarly intervention is the article or the essay; here, the fine, specific points on which scholarly discourse turns can be raised and argued at some length. The 'argument' of any particular intervention is always already part of a much larger argument, of which the implied reader will be fully aware. Many of the 'books' of historical scholarship published at this time, therefore, tend to be compilations of smaller texts – lectures, journal articles and essays – and while individual chapters may perhaps be conceptually coherent, the overall texts tend to lack the formal developmental coherence of typical purpose-written volumes.

As an example of this phenomenon, consider two books published within a few years of each other at the beginning of our period. Robin Flower was an English scholar steeped in the debates on Irish literary history. His book *The Irish Tradition* (1947) is in fact a posthumous collection of various lectures and papers from as long ago as 1926, or as his widow writes in the preface, 'a selection he had put together of what he had already said or written on the subject on various occasions over a long period of years'. There is little or no narrative cohesion in the text. Instead, it relies for impact on the usual tactics of scholarly discourse: the promise of a final truth, perfectly encapsulated in the title, and a register of prestigious fellows – David Greene, Eleanor Knott, Gerard Murphy, Myles Dillon, D.A. Binchy – which evokes the power and prestige of the whole scholarly community.

T.R. Henn, on the other hand, although Sligo-born, was very much part of a British critical tradition and quickly became part of the new international movement in Irish-related literary criticism. His book *The Lonely Tower: Studies in the Poetry of W.B. Yeats* (1950), although indicating a compartmentalised approach – 'Studies' – was written to a thesis, and this thesis (roughly, a delineation of Yeats's complex response to the modern world through the 'key' of ancestral memory and the use of the mask) informs all the individual readings. Each chapter stands on its own but at the same time is part of a larger organic whole – the book – which is greater than the sum of its parts.

The greater length of the book form compared to the essay or the article was not, therefore, as significant for historical scholarship as it was for other critical practices. Although the longer format did allow the writer to develop an argument at length and activate the reader's creative faculties (as with critical books on modern literature), that same reader, because of the specialised subject matter, would generally have been familiar with the range and procedures of the discourse, and consequently less likely to allow formal techniques to influence conceptual considerations. There was no room, in other words, for a macro-thesis which would challenge the legion of micro-theses. And for this reason, historical scholarly discourse, the state which nurtured it and the national identity on which both traded, all remained locked in a mutually sustaining, almost unassailable embrace.

It sometimes happened that an individual scholar coveted and borrowed some of the affective properties from a different discipline. Robert Farren, for example, was a distinguished Dublin-born poet, philosopher, playwright and all-round man of letters. The purpose of his volume of literary history, *The Course of Irish Verse in English* (1948), is to show 'the course of Irish poetry in the English language; to observe and remark upon the growth in Irishness, in separate existence from English poetry, of the poetry that was and is composed in Ireland or by Irishmen' (xi). Just like the late Robin Flower, that is, Farren is in search of 'the Irish tradition', something which already exists and only needs to be hunted down employing the sanctioned scholarly tools. In other words, Farren has borrowed from historical scholarship the search for truth, which is the central, though silent, principle on which that discourse turns. There *is* an Irish tradition of English-language poetry; once identified and described by the scholar-critic, individual examples can be examined and judged on the degree of their adherence to, or divergence from, that tradition. In Farren's book, for example, Goldsmith has 'distinctively Irish moments' (1); Swift has 'decidedly Irish features' (3); Thomas Moore fights (albeit unknowingly) against the 'cloying symmetry' and 'harmony' of English verse using the 'delightful asymmetries' and 'dissonance' of Gaelic 'Amhran' (song) which he had apparently picked up accidentally just by being born in Ireland (6–9). Revealingly, as in the last example, the Irish tradition as identified by Farren is a naturally occurring antidote to the 'too perfect' sophistication of the English language, introducing into that merely cultural discourse the corrective of 'natural shapes, human, animal, still' (6).[8] A natural image is also employed to portray the correct relations between different national traditions:

> It is to be hoped that Irish poets will never cease to read the poetry of other countries and to absorb and adopt to their own ends whatever other countries have to teach them; but this must be done as the body absorbs and transforms its food – as a means of nourishing a poetry which remains its own – : it is scarcely a sign of rude health if a tiger's body turns into a lion's. *That* would not be in the order of natural events, and the tiger's body could not be said to be finding its greater perfection.
>
> Irish poetry on the whole, and certainly in its better part, is decidedly Irish. (167)

The search for truth, then, which Farren has imported into his own practice from the discourse of historical scholarship, becomes a search for nature, nature as it can be identified in English-language Irish verse. In this way, contemporary specialised scholarship (truth), nineteenth-century Romanticism (nature, especially in the Teutonic variety) and Arnoldian cultural politics (Celtic magic) are combined in Farren's discourse to produce a model of Irish literary history which is in complete accord with the principles of radical decolonisation and its insistence on strongly defined, naturally occurring national identities. This emerges most revealingly in the last sentence of the above extract which, in its unproblematic oscillation between descriptive and evaluative uses of the word 'Irish', is a perfect example of radical decolonisation appropriating a discourse – in this case, literary history – and activating it in the name of a discrete, pristine national identity.

On the other hand, a writer working within the tradition of professional historical scholarship might desire the reader-engaging capacities which came with writing a more coherent thesis-led book. In his book *Early Irish Literature* (1948) Myles Dillon is just such an author. The text begins with the usual preface in which the problematic is defined, sources are cited and colleagues acknowledged. The purpose of the book, he writes, is 'to present the imaginative literature of Ireland in a coherent order, choosing only the best that has survived, so that a wider public may become familiar with its content and with its forms' (v). Something like this was to be attempted on a larger scale with the commissioning of the Cultural Relations Committee series a few years later. Here, however, one finds an independent scholar offering readings of individual sections of Irish literary history (the four Cycles, the

Adventurers, the Voyages, the Visions, and Irish poetry) but at the same time presenting an overall interpretation of the whole imaginative tradition. This interpretation is offered at various points throughout the text, but also crucially in the 'Introduction' (xi–xix), a rarity for book-length historical scholarship.

Dillon's introduction is still very obviously working within the limits of his native discipline. Thus we have the specialised languages, the appeal to evidence and research, the scholarly community as tacit reader and the hunt for accuracy (and its hypostasis, truth), encapsulated in phrases such as 'It has long been held' (xi); 'is not convincing' (xi); 'seems genuine' (xii); 'an important source of evidence' (xii); 'no satisfactory account' (xii); 'considerable value' (xvi). At the same time, there is a sense of an overarching idea or argument running throughout the Introduction and the text itself, besides the idea of an Irish tradition which is central to the discourse. This argument is loosely based on the materiality of Irish texts and the many and diverse contexts and forms through which they pass before the scholar confronts them. Thus a developmental register is introduced into the discourse, one that operates at the level of the subject (the narrative of the increasing availability of Irish imaginative literature through literacy, copying, print and so on) and at the formal level (the author's access to, and employment of, versions of this literature, be they originals, manuscript or print copies, facsimile or some other form). It is a small concession to the book-form itself, an acknowledgement by Dillon of the powers of narrative which simultaneously introduces an element of self-consciousness into his discourse. The reader confronts more than just a series of independent readings, but is offered one coherent narrative – 'the Irish tradition' – and a sub-narrative which reflects on the emergence of that tradition.

These are some of the ways in which professional historical scholarship operated in Irish cultural discourse in the period between 1948 and 1958. In the face of growing international influence in Irish cultural debate, monographs needed to be produced. But the book-form itself presented various problems for a discipline which depended on minor interventions into an all-encompassing, though seldom-enunciated, masternarrative. At the same time, the primary search for truth which animated scholarly discourse could serve as a model for critics from outside the scholarly domain. In spite of some concessions to the narrative propensities of the book-form, however, historical scholarship at this time continued to fetishise method as the royal road to the truth of the Irish tradition. This fact, when combined with its unselfconscious institutional- and state-affiliations, meant that the discourse of professional specialised scholarship in Ireland in the 1950s functioned as a major intellectual plank in the continuing dominance of radical decolonisation and its disabling politics of identity.

An Irish Tradition?

There were in fact very few Irish critics writing longer works of literary criticism or history during the 1950s. Because many of these individuals were also creative writers there was less time to devote to a practice which called for what appeared to be a completely different order of intellectual activity. Many also had full-time jobs as teachers or journalists or civil servants which would have made it difficult to find the time or energy for sustained critical thought. Energy, in fact, was the *second* most important problem for the Irish critic at this time: if one's constant theme was the lethargy into which the nation had slipped, it would be difficult to prevent a certain amount of spiritual and intellectual enervation infiltrating one's own work. Content, that is, was constantly threatening form, as the general economic and cultural malaise noted by the critics registered in critical discourse as an inability to formulate a sustained and effective critique of the contemporary situation.

The exception which threatens to prove this rule was provided, as with so much else in Irish cultural debate at this time, by Seán O'Faoláin in his book *The Irish*, first published in 1947. In this volume, O'Faoláin attempted to deal with the historical roots of the Celts, the nature of the conflict with England and some of the main features of contemporary Irish life, including a chapter on 'The Writers'. Even this effort, however, can be seen not as a spontaneous domestic twitching of the critical nerve but a response on the part of an important international writer to external pressures on Irish cultural debate. O'Faoláin's rationale in this text is gleaned from the thought of the Cambridge philosopher of history R.G. Collingwood, and the question which he looks to address throughout *The Irish* is: 'What has this event or this contributed – with whatever racial colouring is no matter – to the sum of world-civilization?' (1980: 10). In other words, O'Faoláin can only formulate his examination of Irish history in terms of a universalist narrative of 'world-civilization', a gesture which engages with (and thus reinforces) the boundaries between 'racial' and non-racial experience.

This point leads on to what was (as remarked above) the single most important influence on Irish critical discourse at this time – the coming of the international (but especially American) critics. With this development, an enormous amount of pressure was generated within the discourse itself as both the literary and critical boundaries of Irishness were challenged systematically for the first time since the cultural renaissance at the turn of the century. This challenge, which operated at both the conceptual and the formal level, simultaneously reinforced and undermined traditional systems of thought. 'Irishness' was recognised internationally as a unique and discrete order of experience; at the same time, the critics' political affiliations, the form of their judgements as well as the audience for which they wrote, were no longer stable points in the Irish cultural equation. This was because any American or European academic, as was increasingly obvious, was free to bring the fruits of modern critical thought to bear on 'Irish' literature, thereby producing an interpretation which would shatter that 'unique' order of experience and its parasite identities. In the light of these developments, the question facing the Irish critics of the 1950s was whether it was in fact still possible for them to engage in Irish-related critical discourse at all, and if so, what form that engagement could take.

One response available to the Irish critics of this time was to attempt to compete with the foreign critics on their own terms. This meant the writing of generally positive, thesis-led books – monographs, critical biographies, genre-histories or extended conceptual essays. If it was impossible to reclaim the realm of Irish-related literary criticism in racialist terms, it might still be possible to maintain an influence in the discourse by aping the techniques and styles of the invaders.

Two texts which can be seen as emerging from this response are *Poor Scholar: A Study of the Works and Days of William Carleton* (1947) and *Modern Irish Fiction: A Critique* (1950) by the writer-critic Benedict Kiely. A regular contributor to domestic cultural debate and a frequent visitor to America, Kiely could not have missed the changes that were overtaking contemporary Irish-related criticism. Rather, it appears that his exposure to the growing international interest in Irish literature left him well disposed to learn its ways and practice them himself.

Poor Scholar is a standard literary biography in that the critic picks a reasonably well-known but (at the time) little-researched figure from Irish literary history, and in the course of an extended analysis of the figure's life and work re-arranges the Irish literary canon to make room for this new 'individual talent' while simultaneously confirming the existence of the tradition formed by that canon. Carleton, according to Kiely, is 'the greatest novelist that Ireland in the nineteenth century gave to the English language' (1972: v), and this book, typically of the new ethos of Irish-related literary criticism, is a celebration of that 'fact'. Although broaching the international perspective that was becoming dominant in Irish-related criticism at this time, therefore, Kiely's discourse is still operating within the traditional parameters of decolonisation, still looking to confirm an Irish tradition and a symbiotic relationship between culture and nation. In this his first attempt at extended literary criticism, Kiely was in fact still coming to terms with the new times in Irish criticism. The text is sparsely annotated, containing no indices or bibliography, and in spite of the flyleaf claims to be 'scholarly if concise', the analysis seldom extends beyond tautologous encomium.

By the time he came to write *Modern Irish Fiction* Kiely had begun to find his critical voice and was happier dealing with the potential incongruities of Irish literature and universalist criticism. He had learned, that is, more of the techniques of the international critical trade and was now ready to compete with international critics on their own terms, speaking their own languages with their own accents. Thus, the text is more ambitious in concept and more scholarly in execution, fully annotated and including an index, biographical and bibliographical notes. Also, and typically of the new ethos, the text is offered by the author as the 'first road' (160) across the uncharted territory of modern Irish fiction. To extend the metaphor, there is a sense throughout the volume of Kiely trying to get in first on what would soon be an incredibly fertile field of research. In other words, the book is written with an eye to the market and to the critic's own career, and in all these ways Kiely tries to preserve a place for himself in a discourse to which he no longer necessarily has access by geographical consideration.

At the same time, Kiely still finds it difficult to escape the old politics of truth which demanded the critic's allegiance to one or another strategy of decolonisation. In the 'Foreword' to *Modern Irish Fiction* he writes:

> Yet the struggle for national independence and distinctiveness has coloured the whole development of Irish literature written in English, and the removal of the necessity for that struggle was bound to affect writers – as it was also bound to affect civil servants, businessmen, and farmers ... It is my hope that the following pages will at least show that the stories did not end when the struggle ended, and that the Irish prose fiction of the last thirty years – probably more important than Irish poetry

> or Irish theatre during the same period – opens a wide window into the soul of the people of this island. (vi–xii)

Here, Kiely claims a key indicative role for literature in Irish history and a key function for his own deciphering discourse, and this strategy is typical of modern international criticism. Yet that literature and that history cannot be formally separated, as one is 'bound to affect' the other, while both are crucially linked to 'the soul ...' (that is, the irreducible identity) ' ... of the people of this island'. For expedient reasons, Kiely bends the knee to modern international criticism; yet he cannot escape the older orthodoxy, the one which claims his discourse as an intervention in the ongoing debate over Irish identity and the relationship with the former colonial power.

There would appear, then, to have been a problem at this time for the Irish critic who adopted international critical techniques. Whereas the foreign status of the American or British critic could potentially undermine the concept of a unique local identity, the national critic could only confirm the discrepancy between her critical practice and literary subject, thus slipping back into the struggle between two already-existing narratives of decolonisation. So, whereas a foreign critic employing foreign methods to talk about Irish literature might unsettle those narratives, an Irish critic employing foreign methods to talk about Irish literature could be re-absorbed unproblematically into a narrative of difference-versus-equality. Thus, the outmoded choice between liberal and radical decolonisation, might be once again confirmed.

Kiely's affiliations were first and foremost literary, and the same is true of most other Irish book-writing intellectuals of the 1950s, including the two yet to be looked at in this chapter, Seán O'Faoláin and Arland Ussher. In the previous generation these critics would have pronounced on matters of national identity with authority and, more importantly, impact. Having rejected the effete intellectualism of the historical scholars and having likewise resisted the relative security of state-affiliation, these writer-critics found themselves on the margins of postcolonial Irish life. The domestic audience for their work was still fairly restricted, and the country's few professional writers, such as Seán O'Faoláin and Frank O'Connor, increasingly had to turn away from Ireland to find subjects for their work and new critical languages with which to speak of these subjects.

O'Faoláin, as mentioned in earlier pages, is perhaps the major Irish critical figure of the period. It was he more than any other individual who kept the notion of an alternative national identity alive through the bleak early years of the post-revolutionary period. He was also one of the few Irish writers to be able to reverse the trend which saw international critics dominating Irish literature. In his books *The Short Story* (1948) and *The Vanishing Hero: Studies in the Novelists of the Twenties* (1956), O'Faoláin writes with authority on matters of international literary interest, reserving only a small section of each volume for analyses of Irish literature.

The two books are typical of contemporary critical practice in that they are thesis-led, genre-based studies. The first is an analysis of the development of the short story as a literary form in which the critic employs a wide range of primary and secondary reading to describe the matrix of personal and technical factors from which the genre emerged. The text includes individual studies on Russian, French and American writers, and technical analyses of conventions, subject matter, construction and language. The critic's credentials, therefore, and his right and ability to intervene in this area of intellectual inquiry are displayed for all to see. Refusing to be restricted by his nationality – indeed, responding to the example of international criticism's attitude to Irish literature – O'Faoláin takes the whole world as his text, thereby giving the lie to the concept of authentic experience as key to critical insight.

Within this general narrative there is time for some local observations, as when O'Faoláin notes that 'Irish writers are far less interested in the technique of writing than in the conditions of writing, though inclined to think exclusively in terms of their own local conditions and to imagine them unique' (23). Such a remark registers unproblematically in terms of an ongoing Irish cultural debate; at the same time, the point is merely one node in a larger, more significant network – 'the short story' and its general, universal problems. For example, in the struggle between natural talent and learned technique (imagined in terms of an extended metaphor of water supply), O'Faoláin writes: 'Surely, if there ever was a writer whose painful efforts to manage his own nature and circumstances makes nonsense of this Divine Plumber theory, it is George Moore' (30). Moore's significance, that is, lies not in his impact on Irish literary practice (although that is described in some detail) but in terms of the larger issue of the relationship between creative imagination and formal technique. Moore's local practice, therefore, is incorporated into a narrative of universal literary practice and in this way Irish experience is raised to a position of equality with the experience of other nations. In other words, in spite of his efforts to take the longer, larger view and thus escape an embarrassingly outdated local debate, O'Faoláin's discourse can be reclaimed by that debate as a typical tactic of liberal decolonisation and its insistence on the parity between discretely defined national identities.

This is also true of *The Vanishing Hero*, O'Faoláin's 1956 study of the modern novel. He cannot write a book on such a subject without mentioning the work of James Joyce; to do so would undermine his thesis, as Joyce's impact on modern fiction was considered seminal by contemporary critical opinion. The great Irish novelist must be considered, if only to be rejected. But he is also constrained to deal with Joyce in terms of the local debate about national identity, as to ignore him might be construed by his opponents in that debate as a statement of opinion. O'Faoláin was in fact trying to maintain an Irish perspective in his critical discourse, but show it in relation to a larger picture, and this tactic perfectly mirrors the kind of Irish identity he argued for throughout the post-revolutionary period.

Again, the critic can make remarks and observations that would have a special significance for an Irish reader familiar with local debates on the relationship between literature and national identity. Thus, when discussing the writer's relationship with his milieu, he writes:

> This idea or notion of fitness is purely personal to the writer if he is an individual writer, and all good modern writing has, perforce, to be individual: i.e. it does not any longer present traditional or social ideas of what is fitting ... It is therefore essential to be aware of the constituents of Joyce's, or any writer's, way of looking at life. (211–12)

There then follows a close analysis of Joyce as individual – his temperament, his religious and political outlook, his personal history. It is, in other words, an author-centred critique, the point being that 'Joyce's world is not Dublin and all that. It is as private a world as that of Proust' (211). This critical attitude flies in the face of the nationalist orthodoxy, so familiar to O'Faoláin through his battles of the 1930s and 1940s, which argued for the interdependence of the individual writer and the community. This point is then borne out by the critical text taken as a whole, which sees Joyce's representation of Irish life studied alongside the work of English and American novelists. The 'vanishing hero' of modern fiction is a larger concept than Joyce, although he emerges from the text as its foremost exemplar. Joyce's work is incorporated into an overarching narrative of the development of modern fiction, and the local experience though which he realises his representative artistic vision emerges from O'Faoláin's discourse as an interesting, but ultimately incidental, feature.

Like Kiely, then, the juxtaposition of 'foreign' elements (the book-form, the subject, the critical methodology) and 'local' elements (Seán O'Faoláin as famous Irish critic, the work of Moore, Joyce, etc. as part of the thesis) threatens to undermine traditional decolonising ideology; like Kiely again, however, the inability or unwillingness of O'Faoláin to displace the opposition between 'foreign' and 'local' experience ultimately means that his work can be reclaimed by that ideology. The difference between the two critics is that whereas Kiely's discourse could not avoid stressing the (radical) difference between Irish and non-Irish experience, O'Faoláin's discourse could be (and subsequently has been) reclaimed in terms of the liberal strand of decolonising thought which looked to demonstrate the similarities between Irish and non-Irish experience.

The political, cultural and critical categories of decolonisation also inform the two Irish-related books written during this period by the philosopher and man-of-letters Arland Ussher: *The Face and Mind of Ireland* (1949) and *Three Great Irishmen: Shaw, Yeats, Joyce* (1953). Unlike Kiely and O'Faoláin, however, Ussher employs a number of techniques and strategies in these books to problematise these categories, threatening to fracture the discursive boundaries between Irish and non-Irish experience and introduce a measure of ironic displacement into the narrative of Irish decolonisation.

In the first place, the material with which Ussher deals (especially in the later book) is in itself problematical. Shaw the socialist, Yeats the Protestant, Joyce the exile – in the cultural climate of post-colonial Ireland the evaluative ('great') and descriptive ('Irishmen') status of these three writers was far from certain. The title of the latter book is thus a provocation to those who defined Irishness in narrow political, religious and domiciliary terms.[9] Moreover, all three demonstrate (although Joyce would appear to be the first among equals) the 'essential Irish quality' which Ussher defines paradoxically 'as a combination of mysticism and irony'. This is not, however, the colourful, exotic mysticism, nor the gay, fatalistic irony of Arnoldian Celticism; instead it is a serious philosophical and political turn of mind which is traced by Ussher in *The Face and Mind of Ireland* through Irish cultural history in figures like Wilde, Swift and Berkeley:

> The Irishman, as I see him, is something of a realist and something of a mystic. In his literature he wavers continually between fantasy and

> farcicality; his most successful genre – from the Cuchulainn epic to *Ulysses* – is a sort of surrealistic extravaganza which has no precise parallel elsewhere. (32)[10]

Against contemporary decolonising thought and its insistence on rigidly defined national identities, Ussher offers a definition of Irishness ('no precise parallel elsewhere') which in its formulation – mysticism and irony, fantasy and farcicality – problematises the very concept of a 'definition'. Shaw, Yeats and Joyce are 'great' to the extent that they exemplify this continual wavering 'between fantasy and farcicality'; but for Ussher it is not the extremes of fantasy and farcicality, mysticism and irony, that define Irishness, but the wavering itself, the constant willingness to travel between categories and essences, experiences and identities. 'Irishness', then, the ostensible subject-matter of these books, was a problematic formulation, because as defined by Ussher it contained within itself the seeds of its own destruction.

This assault on the concept of categorical national identity was intensified by Ussher's uncertain authorial status. Although 'an Irishman by birth and ... by choice also', he insisted that he has 'never been able to associate myself completely with any Irish or Anglo-Irish group' (1949, 'Foreword'). He did not attempt to locate himself outside the question of nationality by shying away from those qualities and experiences – lineage, language, religion, cultural tradition, etc. – which constituted the currency of the debate, but neither did he accept this currency as unproblematic. An Irishman with English blood, a Protestant student of Catholicism, a Gaelic scholar with no ties to the revivalist movement, an Anglo-Irish landlord with no regard for the Anglo-Irish community – Ussher was a living contradiction whose existence gave the lie to that system of thought which operated solely in terms of sharply defined national categories. The coming to dominance of 'foreign' critics in Irish cultural discourse was not, therefore, a cause for concern for Ussher, as he was already both 'foreign' and 'Irish' – indeed, his discourse depended upon the ability to move between these categories, viewing and criticising Irish literature now from one perspective, now from another. This brings us once again to the question of method.

In the preface to *Three Great Irishmen* Ussher writes:

> I use here again the method which, in *The Face and Mind of Ireland*, I called 'the continually shifting viewpoint'. It is not merely that one likes and dislikes, but that one praises and censures for partly the same reasons. This 'ambivalent' approach should need no apology in the age

> of psychoanalysis and relativity; and it would seem especially well adapted to the study of three characters who, whatever their divergences, were all of them masters of irony – even of 'romantic' irony. Nevertheless it is an approach which is still apt to disconcert when employed in philosophy and literary criticism (and philosophy is properly criticism applied to existence itself as art-object). (1968: 10–11)

Such passages (so suggestive of the system that will become known as 'deconstruction') provide Ussher with the basis for what might be termed a 'partial' engagement with received decolonising discourses – 'partial' in the sense of being self-consciously only 'part' of the story while also being inherently biased ('partial') in terms of particular spatial, temporal and discursive factors bearing on the analysis. The 'continually shifting' method complements both the problematic subject-matter and the text's uncertain authorial status. A philosopher by education, Ussher is here engaging in a form of literary criticism, in the process disrupting the boundaries between these traditionally discrete disciplines. He is discussing 'literary' phenomena – Shaw's drama, Yeats's poetry, Joyce's fiction – but not in any straightforward literary-critical way. In fact, the range of discourses employed to discuss the material includes practical criticism, politics, history and metaphysics, and the reader is having continually to reposition herself in response to these discursive shifts. Ussher acknowledges the advent of new times in Irish-related criticism, but refuses to adhere to the agenda being set by what he scornfully refers to as 'the Joyce-cult of the moment' (10). There is nothing overtly scholarly about the discourse – no bibliography or indices, footnotes kept to a minimum – yet it is raised above mere *bellettrism* by the author's adherence to a central thesis and his employment of a wide range of literary and philosophical sources in support of that thesis. Between them, he claims, Shaw, Yeats and Joyce offer 'something like a complete truth by which a man can live' (11); but this truth does not exist outside the critic's own discourse or his very specific activation of the work of these three writers. The critical text, that is, is acknowledged as a text, as an event in its own right; it has an argument, but does not try to conceal the contingent nature of that argument by appealing to the 'evidence' of the 'primary' text. *Three Great Irishmen* is a self-conscious production of Shaw, Yeats and Joyce in which Ussher offers an interpretation of the work of these three writers in its relationship with, and construction of, Irish national identity. Both the interpretation and the work, however, are produced in such a way as to reflect ironically on the very concept of national identity.

In terms of content, authorship and methodology, therefore, Ussher confronts and problematises the received systems of decolonising thought. While continuing to deal with issues of national identity, he refuses to consider that debate on its own terms, and this disruptive strategy of simultaneous acceptance and refusal points to those modes of critical engagement which might signal the onset of a genuine postcolonial politics of displacement.

Ussher is, if anything, the exception who proves the rule. Irish book-length literary criticism in the 1950s remained to all intents and purposes trapped within the terms of traditional decolonisation. Whether its impetus was state-affiliated aggrandisement, as with the texts written and published under the auspices of the Cultural Relations Committee; whether adhering to the codes and practices of professional scholarship and its drive towards truth; or whether reacting to the increasing international critical interest in Irish literature and the growth in the number of monographs, Irish literary criticism effectively failed to imagine a new agenda, or construct a new language, for its debates.

Notes

Introduction

1. See the introductions to Deane, 1990: 3–19; Foster, 1993; xi–xvii; Longley, 1994: 9–68; Gibbons, 1996: 3–22; Lloyd, 1993: 1–12; McCormack, 1994: 1–27.
2. On the question of introductions and prefaces see Spivak's introduction to her translation of Derrida's *Of Grammatology*, 1976: ix–lxxxvii.

Chapter 1

1. This question has emerged in Irish Studies in recent years as a general debate about historical revisionism and also as a series of specific encounters on *The Field Day Anthology of Irish Writing*, 1991. See Brady, 1994; Deane, 1990: 3–19; Foster, 1986; Gibbons, 1991, 1994; Longley, 1994; Mulhern, 1993.
2. For pejorative estimations of nationalism as an invented tradition see Gellner, 1983; Hobsbawm and Ranger, 1983; Kedourie, 1960.
3. On the divided nature of the European Enlightenment tradition and its contradictory inheritance see Adorno and Horkheimer, 1979.
4. See chs 4 ('Nationalism and Cultural Identity') and 5 ('Nations by Design?') in Smith, 1991: 71–98.
5. Moi, 1985: 12. Although the disciplines on which I am focusing throughout this study are predominantly historical and cultural-critical, the use here of terms such as 'the symbolic' invokes areas of research with which I have neither space nor competence to deal. It does seem clear, however, that much modern psychoanalytic discourse has been concerned to revise the predominantly Freudian basis of earlier postcolonial criticism – such as Fanon, 1967, 1986; Mannoni, 1964; and Nandy, 1983 – and to draw instead on post-Freudian psychoanalysis to address the issues of colonial discourse and anti-colonial resistance. Theories such as Lacan's

'imaginary/symbolic' (1977), Kristeva's 'symbolic/semiotic' (1981, 1984), and the 'schizophrenia' explored by Deleuze and Guattari (1977) all appear to offer fertile lines of enquiry, as well as interesting correspondences to the 'liberal/radical' dyad explored throughout this book.

6. Official Irish state-nationalism after 1922 takes the form of 'Eastern nationalism' described by Partha Chatterjee, organised around 'a regeneration of the national culture, adapted to the requirements of progress, but retaining at the same time its distinctiveness' (1986: 2). See Gibbons, 1996: *passim*, but especially ch. 5, 'From Megalith to Megastore: Broadcasting and Irish Culture' (70–81), on the competing claims of state and nation in post-Treaty Ireland.
7. For an elaboration on 'necessity' and 'insufficiency' as coterminous properties of anti-colonial discourse see Terry Eagleton's essay 'Nationalism: Irony and Commitment' in Deane, 1990: 23–42.
8. 'Can the Subaltern Speak?' is the title of an influential essay by Spivak (republished in Williams and Chrisman, 1993: 66–111), and a question to which she appears to reply in the negative.
9. A recurring argument of many poststructuralist theories is that there is no non-ideological position from which to criticise the effects of ideology. Narratives of change are only possible from within established discursive limits – as Derrida writes in an analogous context: '*There is nothing outside of the text*' (1976: 158).
10. This collection of essays is probably the main source for the vogue of hybridity as a critical (and increasingly creative) concept. For an analysis of the historical provenance and contemporary function of the term see Young, 1995.
11. Again, such a strategy finds an echo in Pêcheux' model of subject identification through language, in which he writes: 'In reality, the operation of this "third modality" constitutes a *working* (transformation-displacement) *of the subject-form* and not just its *abolition* ... Ideology ... as the process of interpellation of individuals as subjects – does not disappear, but operates as it were *in reverse*, i.e., *on and against itself*, through the "overthrow-rearrangement" of the complex of ideological formations (and of the discursive formations which are imbricated with them)' (1982: 159, original emphasis).

12. See Graham, 1996, for an attempt to read Maria Edgeworth's *Castle Rackrent* in terms of what Bhabha calls the 'ambivalence at the very origins of colonial authority' (1994: 95).
13. Spivak's investment in literature as an exemplary form of resistance echoes Derrida's insistence that 'It was my preoccupation with literary texts which enabled me to discern the problematic of writing as one of the key factors in the deconstruction of metaphysics ... Literary and poetic language can provide the non-place from which to deconstruct philosophy' ('Deconstruction and the Other', in Kearney, 1995: 156–76 (159, 162)).
14. Many commentators have insisted that hybridity colludes with the reproduction of dominant bourgeois ideology and the late capitalist phase into which it has moved. Celebrations of flux, flexibility and indeterminacy have been criticised as merely a critical rationale for the new phase of global capitalism into which we have moved rather than an explanation of the non-West's mode of resistance and survival. For this and related critiques see Sangari, 1995; David Lloyd, 'Ethnic Cultures, Minority Discourse and the State' in Barker, Hulme and Iversen, 1994: 221–38; Harvey, 1989; Smith, 1991: 157–60.
15. See especially Derrida's *Of Grammatology*. Spivak's eighty-page introduction to her translation of this text constitutes in itself an important source for anglophone poststructuralism.
16. 'Interview with Edward Said' in Sprinker, 1992: 248–9.
17. On Said's engagement with the conflict within postcolonial theory between poststructuralism and conventional historiography, see Bill Schwarz, 'Conquerors of Truth: Reflections of Postcolonial Theory' in Schwarz, 1996: 9–31. W.J. McCormack has been the most strident Irish critic of what he terms 'Weetabix Theory, incredibly dense and regular in structure, but lighter than its box' (1993: x).
18. Derrida himself has commented on both the misuses of his own work and the politically quietistic implications of certain un-named theorists (although Baudrillard and Lyotard would seem to be the principal targets): ' ... the necessary deconstruction of artifactuality should never be allowed to turn into an alibi or an excuse. It must not create an inflation of the image, or be used to neutralise every danger by means of what might be called the trap of the trap, the delusion of delusion: a denial of events, by which everything – even violence and suffering, war and death – is said to be constructed and fictive, and constituted by and for the media, so that nothing really ever

happens, only images, simulacra, and delusions ... the only responsible response is never to give up the task of distinguishing and analysing. And I would also say: never to give up on the Enlightenment ... ' (*Passages*, 1994: 29, 34).

19. See the opening chapters, 'Nationalism as a Problem in the History of Political Ideas' and 'The Thematic and the Problematic' in Chatterjee, 1986: 1–53. This is perhaps the moment to clarify that in identifying different 'modes' of decolonisation I am not advocating a theory of nationalist periodisation, such that postcolonial nations may be 'sympathetically' understood as travelling along the 'true' (Western) road towards 'those universal values of reason, liberty and progress' (Chatterjee, 1986: 6). As Stuart Hall insists (in his essay 'When was "the Post-colonial"? Thinking at the Limit' in Chambers and Curti, 1996: 242–59): 'What "post-colonial" certainly is not is one of those periodisations based on epochal "stages", when everything is reversed at the same moment, all the old relations disappear for ever and entirely new ones come to replace them. Clearly, the disengagement from the colonising process has been a long, drawn-out and differentiated affair, in which the recent post-war movements towards decolonisation figures [sic] as one, but only one, distinctive "moment"' (247). Just so in Irish critical history which reveals a series of exchanges between various modes of engagement with colonialism rather than a progressive narrative of decolonisation. The most engaging modular theories addressed in the latter part of this chapter (as well as the most original and challenging critical practices addressed in Part 2 of this book) are those which have been concerned to expose the complicity of theories of derivativeness whereby neo-colonial effects are systematically insinuated back into postcolonial societies, and to reveal the extent to which Western systems of thought were necessarily altered in the confrontation with alternative, non- or part-Western systems.
20. Derrida has also exposed the abuses to which the issue of derivativeness is prey: 'How can we make progress if every political critique, every historical re-interpretation, is going to be automatically associated with negationist-revisionism, if every question about the past, or more generally about the constitution of truth in history, is going to be accused of paving the way for revisionism? ... What a victory for dogmatisms it will be, if prosecutors are constantly getting to their feet to make accusations of complicity with the enemy against anyone who

tries to raise new questions, to disturb stereotypes and good consciences, and to complicate or re-work, for a changed situation, the discourse of the left, or the analysis of racism or anti-semitism' (*Passages*, 1994: 34).

21. On Lloyd's engagement with Bhabha, and for his critique of the 'narrative of representation' in Irish cultural discourse, see the essays 'Adulteration and the Nation' in Lloyd, 1993: 88–124; and 'Ethnic Cultures, Minority Discourse and the State' in Barker, Hulme and Iversen, 1994: 221–38.
22. This is the version of nationalism as an 'imagined community' explored by Benedict Anderson. Nationalism, he claims, was made possible by the same advances in print capitalism which saw the rise of modern cultural forms such as the newspaper and the novel (1991: 30ff.).
23. This is a point overlooked by Kevin Barry (1996) in his critique of Gibbons's use of allegory as a figure of resistance. Gibbons does not appear to be arguing 'that a certain kind of aesthetic *corresponds* to a certain kind of politics; that there is one aesthetic of recalcitrance and another of the nation-state, one aesthetic for the oppressed and another for those who have power' (8, emphasis added), but rather that allegory makes itself available as a strategic, and peculiarly useful, form of representation for those on the margins of power. This is not to say that an established, institutionalised discourse such as dominant nationalism might not wish to sample the *specific aesthetic effects* of allegorical discourse for *specific political ends* – say, for example, to achieve the effect of the organic community in an increasingly bureaucratic state.

Chapter 2

1. Besides Said, writers such as Anderson, 1991, Giddens, 1987, and Smith, 1991, have stressed the reliance of state development upon what Giddens calls 'modes of discourse which constitutively shape what state power is' (209). The argument of this chapter and this book is that in the cultural sphere, and in the context of modern anti-colonial state development, criticism is one such mode of discourse, but a crucial one that needs to be rehabilitated from its traditional secondary status.
2. In 'The Order of Discourse' (1970), republished in Young, 1981: 48–77, Foucault writes: 'and since, as history constantly teaches us, discourse is not simply that which translates

struggles or systems of domination, but is the thing for which and by which there is struggle, discourse is the power which is to be seized' (52–3).

3. Such a model is implicit also in Smith's conception of the 'vital role' (1991: 94) played by culture in the emergence of modern nationalism. In a manoeuvre typical of the intellectual framework within which he is working, Smith posits a familiar hierarchy of cultural activity: (1) intellectuals (artists and producers); (2) professional/interpreters (critical intelligentsia) who transmit and disseminate those ideas and creations; (3) a widely dispersed, educated audience/public who consume ideas and works of art (92ff.). This begs two important questions: (1) Where did the 'artists and producers' discover the ideas on national culture which they subsequently committed to artistic form? (2) In what discursive mode did they set about creating a consumer context in which such ideas might have purchase? If, as I am suggesting, the answer to both questions is: criticism – that is, the subjects and activities grouped under section 2) – then that hierarchy collapses, and the established model of the culture/politics nexus breaks down.
4. This project has already been instigated in Irish Studies by critics such as Kearney, Lloyd and Gibbons, as their works reveal a willingness not only to re-evaluate a range of 'primary' cultural materials but also to (self-) reflect upon the critical processes through which such materials have been organised. For the beginnings of a systematic engagement with the critical production of modern Irish culture and identity see Leerssen, 1996a and 1996b; this latter text seems especially suggestive but appeared too late for me to engage with it in any great detail in this project.
5. Williams, 1983: 84–6. Invoking an alternative (Althusserian) theoretical tradition, we might say that criticism is an emblematic form of the modern cultural Ideological State Apparatus in that it functions 'massively and predominantly *by ideology*' but also 'by repression, even if ultimately, but only ultimately, this is very attenuated and concealed, even symbolic' (from the essay 'Ideology and Ideological State Apparatuses', 1971: 127–86 [145]).
6. Eagleton's version of the relations between literature and literary criticism in modern England is contested by Simon During in his essay 'Literature – Nationalism's Other? The Case for Revision' in Bhabha, 1990: 138–53, in which he writes:

'It is becoming a commonplace that the institution of literature works to nationalist ends ... I want to draw attention to the ways that, at least since the founding of the modern state, literature has operated in different social spaces than nationalism, employing different signifying practices. Furthermore I want to argue that literary criticism canonizes those texts which do not simply legitimate nationhood' (138); and later: 'Literary criticism enters the project of preservation which works for subjectivity but against both individualism and theory. As such it shares the non-nationalist politics which inform the objects of its approval, maintaining them in the canon' (149). During's 'literary criticism', as he readily admits, 'is not literary history, not interpretation, not theory, not reviewing. Developed by Eliot and Richards, it is an Anglo-American phenomenon with roots in the German moment of Kant and Schiller' (149). In the present context, three points are worth making in response to this otherwise brilliant thesis. (a) While it certainly seems true that one institutionalised version of English literary criticism adumbrated a 'retreat of the ethical' (153) in its affiliation to a Burkean, counter-revolutionary discourse, it did so within a specifically empowered politico-cultural framework, one relying almost entirely on (English) nationalist assumptions. (b) Such a narrow definition of literary criticism robs During's thesis of much of its force. The principle of commentary saturated modern English society, feeding in many significant ways into the reproduction of that society in its specifically nationalist sense of itself, and to focus upon one privileged model seems both methodologically and politically questionable. (c) As we might expect, versions of, and reactions against, these positions emerge in Irish critical history – it is possible, for example, to trace the development of a critical discourse (pedagogical and *belletristic* in impulse) repugnant to any notion of a link between nationalism and literature. As in proviso (a) above, however, in each case the critical discourse in question is located within a specifically political frame of reference; in the case of modern Ireland, the relevant political framework is that of decolonisation.

7. Despite their many real differences, Derrida and Foucault share a remarkably similar line on the concept of commentary or criticism. In 'The Order of Discourse' (1981), Foucault writes: 'By a paradox which it always displaces but never escapes, the commentary must say for the first time what had, nonetheless, already been said, and must tirelessly repeat what had, however,

never been said ... It allows us to say something other that the text itself, but on condition that it is the text itself which is said, and in a sense, completed' (58). This accords with Derrida's approach in *Of Grammatology* (1996) where he traces the paradoxical relations between sign (literature) and supplement (criticism) – 'supplement' in the sense of something extra and in the sense of making complete: 'Yet if reading must not be content with doubling the text, it cannot legitimately transgress the text towards something other than it, towards a referent (a reality that is metaphysical, historical, psychobiographical, etc.) or towards a signified outside the text whose content could take place outside of language, that is to say, in the sense that we give the word here, outside of writing in general' (158). For a somewhat dated though still useful comparison of the approaches to the problems of textuality of these two seminal modern thinkers see the essay 'Criticism Between Culture and System', in Said, 1991: 178–225.

8. The root of 'criticism' is in the Greek *krino*, meaning to choose or to decide (Williams, 1984: 84–6; Smith, 1988: 43). For Derrida, as Smith goes on to point out, 'criticism or critique embodies a moment of crisis at which meaning is decided. In Derrida's characterization of it, the moment of criticism in the logocentric tradition is critical and decisive because it is the point where metaphysical operations of exclusion takes place, and where the concomitant institution of reason take place. Criticism is thus, in Derrida's pun, a process of *arraisonnement*, a mode of bringing to reason or of righting (as one would right a ship). Insofar as it proposes and practices a dismantling of that kind of exclusive and exclusionary reason, Derrida's notion of critical interpretation is an attempt to remove from critical practice, broadly understood, exactly its moment of critique' (43–4). This appears to be related to the project of *Blindness and Insight* in which de Man is concerned to expose the 'irony' or 'error' upon which western notions of textuality and interpretation are founded, and the 'pure nothingness' underlying the notion of a human subjectivity, which poetic language names 'with ever-renewed understanding' (1983: 19, 18). It is in the context of an exposure of reason's inherent crisis that Derrida and de Man are most frequently linked, and that the latter often is characterised as a 'deconstructionist'. It is far from clear, however, whether the ubiquitous textual and ontological crisis discerned by Derrida and de Man is (or is intended to be) similarly deployed in terms of the

institutional organisations of power into which discourses such as literature and criticism are locked. The ire of Christopher Norris noted in the previous chapter is particularly directed towards variations on what he understands to be de Man's politically quietistic version of deconstruction (1994: 39ff.), a pattern of antithetical thinking he characterises as 'reactive, sterile [and] deadlocked' (72). See also Derrida's *Mémoires: for Paul de Man*, 1986.

9. Eagleton, 1983: 17–54. See also Brian Doyle, 'The Invention of English' in Colls and Dodd, 1986: 89–115, in which he writes: 'In Britain ... English has functioned to provide a substitute for any "theory" of the national life in the form of an imponderable base from which the quality of the national life can be assessed. The sense of "Englishness" that English has come to signify was apparently so free of any narrow patriotism or overtly nationalist or imperialist politics that any debate about the meaning of the term itself seemed unnecessary until quite recently' (111).
10. Part 2 of the present study will be much concerned with the social function of two of the major forms of literary criticism identified by Hohendahl (1982): a professional academic discourse, mostly engaged with the literature of the past 'according to the rigorous rules of scholarship' (14) and mostly located within the university; and an amateur discourse 'closely connected with a mass medium like the press', mostly concerned with the literature of the present and by and large 'unfamiliar with the technical terms of literary analysis and usually uninterested in the professional disputes and disagreements that are an important aspect of academic criticism' (14).
11. In his essay 'Marxism and Popular Fiction' (1981) republished in Humm, Stigant and Widdowson, 1986: 237–65, Bennett had already insisted that any attempt to address 'the function of criticism' should consider the relations between the various kinds of critical discourse that operate in modern western societies. He argued that: ' ... an exclusive focus on the upper echelons of criticism is misleading in that their effects can only be determined if they are placed within a knowledge of the full internal economy of the sphere of criticism, one which includes a knowledge of the way non-canonized texts have been routinely criticized in weekly reviews, the daily press, etc. Nor is this solely required for theoretical reasons. The political value of any critical practice depends on the way it intrudes upon and modifies the relationship between text and reader. It is necessary, therefore,

to consider how that relationship is already constituted: how, in supervising the relations between text and reader, the socially dominant forms of criticism help to structure the ideological field into which the text plays and in which its – always variable – effects are registered' (259).

12. Both Hohendahl and Eagleton base their work on Habermas's notion of 'the public sphere', the former systematically, the latter more loosely. Bennett claims that a similar notion informs much of the neo-Arnoldian debate on 'the function of criticism', including Jameson's *The Political Unconscious* (1981), Lentricchia's *Criticism and Social Change* (1983) and Said's *The World, The Text, and The Critic* (1983).
13. It should be noted that there already exists a branch of sociology which is concerned with what one practitioner (Judith R. Kramer) calls 'the social role of the literary critic'. In an essay of that title (in Albrecht, Barnett and Griff, 1970: 437–53), Kramer undertakes an empirical study of categories such as social background, occupational status, nature of critical works and literary outlets. In 'Mass, Class, and the Reviewer' in the same volume (455–68), Kurt Lang attempts a similar analysis of the 'role of the reviewer whose criticism is regularly featured in the daily press and in weekly publications' (455). Despite their calls for a socially responsible approach to the history of criticism, it is doubtful if Eagleton, Hohendahl or Bennett would embrace a full-blown sociology of criticism, as each in their own way appears keen to engage with textualist and philosophical levels of analysis beyond the remit of quantitative sociology.

Chapter 3

1. See Boyce, 1982; Cairns and Richards, 1988a; Cobban, 1960; McCormack, 1985; McDowell, 1979; Moody and Vaughan, 1986; Vance, 1990. The derivativeness of Anglo-Irish identity at this time is explained by Roy Foster: 'It is, in fact, the British connection that should be stressed: for Irish parliamentary politics continued to react to changes in England, and English politicians continued to see Ireland as an appropriate area for intervention ... the continued derivativeness of much Irish metropolitan culture, the obsession with refining the Irish accent, the provincial snobberies so characteristic of Dublin life, should be seen in this context. The political class

continually emphasized their community of culture and tradition with England; the Gaelic revivalism of O'Conor, O'Halloran and Curry was never really adopted by the "patriot" culture, though figures like Flood had developed antiquarian interests by the late eighteenth century, and country gentlemen competed with each other in their archaeological finds. "Patriot" nationalism remained exclusive: the rule of an enlightened elite, rather than the broadening of national interests that was so important to the self-image of the American revolutionaries' (1989: 251–2).

2. Preston, 'Thoughts on Lyric Poetry' in Anonymous, 1787: 57–73. Preston (1753–1807) was educated at Trinity College. A prolific essayist and minor poet, he was also the first secretary of the RIA. Though politically conservative he supported Catholic emancipation.
3. The volume is comprised of three different sections: Science, Polite Literature and Antiquities. The Polite Literature section includes the following items: (1) 'An Essay on Sublimity of Writing' by the Rev. Richard Stack, D.D. Fellow of Trinity College, Dublin, and M.R.I.A.; (2) 'Essay on the Stile of Doctor Samuel Johnson, No. I' by the Rev. Robert Burrowes, A.M. Fellow of Trinity College, Dublin and M.R.I.A.; (3) 'Ditto, No. II' by the Same; (4) 'Thoughts on Lyric Poetry' by William Preston M.R.I.A. To which is subjoined ... ; (5) 'Irregular Ode to the Moon' by the Same.
4. Mason's long dramatic poem *Caractacus* was published in 1759 and is an early example of the influence of literary medievalism in British poetic discourse. According to Snyder (1923: 53–60) the work went through many drafts under the watchful eye of Gray. Mason edited Gray's papers after his death and it is in a commentary on this work that he had occasion to question the latter's proclivity for the irregular ode. Snyder in fact dates the initiation of the Celtic movement in Britain from the publication of Gray's *The Bard* in 1757, a long narrative ode based on the expulsion of the Welsh bards by Edward I.
5. 'The Royal Society decreed that its members must employ only a plain, concise, and utilitarian prose style suitable to the clear communication of scientific truths ... In polite literature ... the ideal of good prose came to be a clear, simple, and natural style which has the ease and poise of well-bred urbane conversation. This is a social prose, designed for a social age' (S.H. Monk

and L. Lipking, 'The Restoration and the Eighteenth Century' in Abrams, 1979: 1731–2).

6. Walker (1762?–1810) was born in Dublin and was privately educated in that city. He spent some of his early years in Italy for health reasons but returned to Dublin in his early twenties where he became a clerk at the Treasury in Dublin Castle and lived in a large house called St Valeti in Bray, Co. Wicklow, where he maintained a large library. He was a founder member of the RIA and contributed many papers to its 'Transactions'.
7. Memmi, 1957: 19–44. Cairns and Richards (1988a) adapt Memmi's concept to observe that 'conscious rejection does not ensure the erasure of unconscious attitudes and assumptions which frequently surface and reveal the "refuser" to share many of the fundamental assumptions of the class which has been nominally rejected ... one must be alert to the possibility that the moves of the colonizers towards the politics and culture of the colonized are motivated by the desire to achieve influence through an act of association and appropriation rather than identification and (self)absorption' (25).
8. Foster, 1989: 210; Leerssen, 1996a; Snyder, 1923. The Ossianic rage was at its height in the 1780s and was compounded by the controversy over Chatterton's *Rowley* poems. In 1784 the Welsh antiquarian Edward Jones published his influential *Musical and Poetical Relicks of the Welsh Bards*, a volume which was in the possession of Walker and which may have inspired his work and his title.
9. Charles O'Conor the Elder (1710–91) was the descendant of one of the old Gaelic royal families and had managed to hang on to some of his lands during the post-Limerick confiscations. He was one of the few Catholic members of the RIA, and a leading figure in the Catholic Committee which sued for Emancipation at the Irish and British Parliaments. His *Dissertations on the History of Ireland*, including an attack on Macpherson, was published in 1766.
10. In *Irish Bards* Walker makes 549 references to 188 authors from 227 different sources, drawing on material from 38 different literary forms. His main sources are Vallencey (44 references), O'Halloran (41) and O'Conor (37), and the forms he most frequently cites are Histories (56) and Poetry (28).
11. Snyder (1923: 11–12) notes the emergence at this time of a type of crank claiming Celtic precedent for everything valuable in the world, a tendency which Walker himself just about manages to avoid. It was this Celtomania that led one 'scholar'

to claim that the Welsh language was descended from Hebrew, and that the Welsh nation was one of the lost tribes of Israel. Walker's friend and one of his major sources, the English soldier Charles Vallancey, smacks of this tendency, which has obvious affiliations with the development of 'radical' decolonisation in certain extremist directions.

12. On the significance of 'paratextual' discourse see Leerssen, 1996b: 7, 57–64, 242. Even at the time of its publication Walker's research techniques were in doubt, as a highly critical review in the *Monthly Review* of December 1787 (by Dr Charles Burney, a leading English musicologist of the day and one of Walker's main sources) demonstrates: 'So that, beside the difficulty of translating and of ascertaining the antiquity of these poetical Irish witnesses, the Author's materials for filling a large book being scanty, they have been eked out with the dry, formal compliments to friends, and the parade of great reading, displayed in the notes, with the pomp and liberality of a German commentator ... But notwithstanding these innumerable proofs of the Author's acquaintance with books in all the living as well as dead languages, they only remind us that he is a young book-maker, and has not yet read enough to know what has been already often quoted, and what is still worthy of a place in a new book written with taste and elegance.' This last comment must have been especially galling for the Anglo-Irish critic. Walker's reputation did not improve with time; nor were Irish historians kinder to him than the Englishman Burney. Eugene O'Curry, the co-founder of the Irish Archaeological Society in 1840, was of the opinion that 'Walker seems to have been the sport of every pretender to antiquarian knowledge ...' (quoted in Flood, 1905: viii), while Flood himself refers to Walker's work as 'absurd' and 'ridiculous' and shares O'Curry's view of him as a dupe or a charlatan or both. More recent historians, such as Breathnach (1965), while admitting Walker's dilettantism, tend to emphasise the groundbreaking nature of his work, arguing that 'The author of *Historical Memoirs of the Irish Bards* deserves to be remembered for his enthusiastic advocacy of Irish learning when perhaps such advocacy was of greater importance than scholarly competence' (97).
13. On the relations between Anglo-Irish 'Patriotism' of the eighteenth century and subsequent Irish nationalism see Leerssen, 1996b: 14–32. Leerssen's analysis subscribes to a modular history of Irish decolonisation similar to the one

described in this book. He writes, for example, that 'Patriotism is essentially different from nationalism ... The Patriots' agenda invoked arguments of *equity* and just representation of interests rather than an essential, national *difference* between Ireland and England' (20, emphases added).

14. As glossed by Cairns and Richards, Gramsci's notion of hegemony 'takes two main forms: "expansive hegemony", in which the leadership of a particular group advances the interests of all those who are led, and "transformist hegemony", where a particular class or class fraction is benefited, possibly at the expense of the remaining member groups of the class alliance' (1988a: 168).
15. Ferguson's critique is a good example of what Terry Eagleton describes as the 'partisan bias, the vituperation, the dogmatism, the juricial tone, the air of omniscience and finality' with which early nineteenth-century critical discourse conducted its business (1984: 37). Indeed, it is difficult to see how Ferguson might reconcile the virulence of his discourse with the consensus which was his ostensible goal.
16. For Ferguson, as David Lloyd argues, 'Knowledge ... becomes unifying rather than – as with Hardiman – divisive and sectarian. The ideal of transparent translation repeats this theory at two levels. In the first place it ensures a continuous transition into which no arbitrariness enters, rather than furnishing a perpetual reminder of the nation's separation from its past. In the second place, the ideal of tranparency allows for the undistorted reproduction in English of the essential quality of the Gael, which, for Ferguson, as for Arnold some thirty years later, is "sentimentality" ... If Ferguson finds that "their sentiment is pathetic" and that "desire is the essence of that pathos," his mission will be to direct that "desire" away from a lost past and toward the idea of unity' (1987: 84). On Arnold's critical politics as they bear on contemporary colonial relations between England and its 'Celtic' margins see Smyth, 1996a.
17. In the foreword to a collection of Davis's work, Eamon de Valera wrote: 'The opinions and sentiments expressed and the ideals presented are independent of time and condition. They are, I believe, as potent to arouse the enthusiasm and to fire the zeal of the generous-hearted today as they were when the pieces were first written or delivered ... I urge that the essays and poems it contains be read and re-read by our "Young Ireland" until they are known by heart, and have become for them an abiding source of inspiration and an ever-present

incentive to noble action.' Alluding to the Northern state and its unacceptable sequestration from the national trunk, de Valera continues: 'Those who will be inspired by Davis's writings will dedicate themselves to the uncompleted task' (Anonymous, 1945: v–vi).

18. It is in this context that David Lloyd has claimed that 'The Young Ireland movement of the 1840s inaugurated a cultural tradition that conceives the responsibility of literature, and of other cultural forms, to be the production and mediation of a sense of national identity' (1987: xi). Although Young Ireland represents a crucial moment in the normalisation of relations between 'cultural form' and 'national identity', in fact it is possible, as we saw in the previous section, to trace such a relationship in Irish critical discourse back to the late eighteenth century at least. At the same time, Lloyd is surely right when he points out (employing the standard 'derivativeness' line) that Young Ireland's ' ... desire to produce a canonical "major" literature and the consequent adoption of a specific matrix of concepts through which the nature of the canonical is to be defined ironically bring Irish nationalist theoreticians of culture into line with a conception of the canon and, more generally, of aesthetic culture that is intimately linked to the legitimation of imperialism and to that mode of internal political hegemony, liberal democracy, which corresponds within the nation-state to global imperial domination' (3–4).
19. It would be a mistake simply to equate liberal decolonisation with Anglo-Ireland, as many Irish-Irish literary critics embraced this option, especially in the years after Independence, while radical decolonisation found some of its fullest formulations in the work of Anglo-Irish critics such as Davis and Douglas Hyde.
20. 'This passion for perfection seems to me as truly Celtic a thing as the ready indulgence of sentiment; our illuminations, our penmanship, our work in stone and metal, all our arts of design, show an infinite love of taking pains' (99–100).
21. 'Browning' (22 February 1890), quoted in Storey, 1988: 5.
22. For example, Yeats invariably begins his critical essays with an anecdote, a challenge or an exposition of his own critical values: 'When I come over here from London or cross over to London I am always struck afresh ...'; 'I am going to talk a little philosophy ...'; 'The arts have failed ...'; 'When O'Leary died I could not bring myself to go to his funeral ...' (Storey, 1988: 74, 85, 132, 171).

23. The extent to which Yeats can be 'covered' in this way is no doubt problematical. It is an indication of the fluidity of cultural decolonisation that even in his own lifetime Yeats's work could be incorporated into the various strategies of resistance, so that his name functioned as an uncertain signifier within different decolonising modes. The author of *Cathleen ni Houlihan* existed in an uneasy relationship with the castigator of Free State divorce legislation, and even today contextual factors still determine which 'Yeats' is invoked much of the time. In as much as the liberal mode is vital to any narrative of decolonisation, then it is possible to agree with Edward Said's claim that Yeats 'gave us a major international achievement in cultural decolonization' (1993: 288). It is necessary to differentiate, however, between the mature man-of-letters with whom Said largely deals, and the younger writer who in poems such as 'September 1913' and essays such as 'Poetry and Tradition' articulated his resentment at the marginalisation of Anglo-Ireland by the developments of 'radical' decolonisation.
24. Quoted in Duffy, 1894: 161.
25. On the narrowing of Irish identity during the revolutionary period see Brown, 1985: 13–44; Lee, 1989: 1–14; Lyons, 1973: 224–46.
26. Quoted in Mac Aonghusa and Ó Réagáin, 1967: 152. In a passage in *The Wretched of the Earth,* Fanon describes with uncanny accuracy the stage of nationalist resistance in which a figure such as Pearse might emerge: 'The native, after having tried to lose himself in the people and with the people, will on the contrary shake the people. Instead of according the people's lethargy an honoured place in his esteem, he turns himself into an awakener of the people; hence comes a fighting literature, a revolutionary literature, and a national literature. During this phase a great many men and women who up till then would never have thought of producing a literary work, now that they find themselves in exceptional circumstances – in prison, with the Maquis or on the eve of their execution – feel the need to speak to their nation, to compose the sentence which expresses the heart of the people and to become the mouthpiece of a new reality in action' (1967: 179).
27. Mac Aonghusa and Ó Réagáin, 1967: 154. On Pearse's use of Davis see 'History and the Irish Question' in Foster, 1993 (1–20), where he writes: 'Pearse's use of Irish history was that of a calculatedly disingenuous propagandist; it was this that

enabled him, for instance, so thoroughly to misinterpret Thomas Davis' (14).

28. According to this, if Davis himself represented Irish nationalism's *moment of departure* – the attempt to combine 'the superior material qualities of Western cultures with the spiritual greatness of the east' (in this case, of the Celtic margins) – then Yeats represents the *moment of manoeuvre*, consisting 'in the historical consolidation of the "national" by decrying the "modern", the preparation for expanded capitalist production by resort to an ideology of anti-capitalism' (1986: 51–2).
29. Norman Vance argues that throughout the modern period, the main source of resistance to dominant decolonising practices came from the ranks of the Unionist, Anglo-Irish, Protestant population, and that individuals such as Ussher, Dowden and Wilde provided an example for disaffected Catholic intellectuals such as Joyce (1990: 195ff.). This seems too schematic, while also understating the complexity of Joyce's paradoxical relationship with Ireland, characterised by simultaneous and ultimately undecidable discourses of affiliation and alienation.
30. Arnold made his seminal contribution to the debate in 'The Function of Criticism at the Present Time' (1953: 137–64), first published in 1865 as an introduction to *Essays in Criticism*. For Arnold's influence on Yeats, both in terms of subject matter and critical ideology, see Sena, 1980.
31. Terry Eagleton has stressed the inter-relations between Wilde's Irishness and his 'remarkable anticipation of some present-day theory' (1989: viii).
32. Synge, from 'Can we go back into our mother's womb? A Letter to the Gaelic League' in Price, 1982: 400. Moore's focus was on French and Russian fiction in the two volumes of criticism – *Impressions and Opinions* (1914) and *Avowals* (1919) – published after his disenchantment with the cultural revival. The Yeats quote is from 'Sailing to Byzantium' (1928), 1933: 217.
33. There is no space here for a full explication of Joyce's challenge to Irish cultural nationalism, although some of the analyses which have influenced the position he occupies here include MacCabe, 1979; Nolan, 1994; and Eagleton, 1995. Many commentators have remarked the recalcitrance of Joyce's work for traditional critical discourse: 'Joyce's excesses can never be contained: no critical consensus will ever be able to control the rapturous and rupturing laughter provoked, inside and outside the text' (Attridge and Ferrer, 1984: 2). Making a related

point, Seamus Deane argues that because it may be mastered from so many specialised (and thus partial) positions, *Ulysses* constitutes a radical threat to the notion of criticism as a coherent institutional activity, introducing elements of contingency and interdisciplinarity at odds with traditional textual scholarship: '*Ulysses* is a complicated score that can be played in as many ways as there are conductor-experts – the Dublin expert, the Jesuit specialist, experts in the classical references, the literary allusions, the Irish history. Every reader is an expert in Joyce depending on the reader's speciality' (1985: 90).

34. Nolan (1994) surveys the recruitment of Joyce for anti-nationalist and revisionist critical discourses, and offers an analysis which shows that his work 'cannot be enlisted for an aloof, enlightened cosmopolitanism' (48) or any model of a 'weak, constitutional nationalism', but may instead be understood as consorting much more closely than was previously imagined with an extreme form of cultural nationalism which 'was founded on secular principles of universal rights that may make it appear much less narrow and provincial than its "liberal" alternative' (19). Nolan's position might be compared with that of Benita Parry, who asks whether the prospect of a post-nativist 'whole man' promulgated by a postcolonialist criticism essentially opposed to any but the weakest of nationalisms 'is one that wholly delights' (1994: 192).
35. O'Faoláin later revised his opinion of Joyce, although the terms in which he is salvaged still appear misplaced: Joyce is 'our one great realist', a 'writer offering at least one coherent and liberating idea to the generation emerging after Yeats', while O'Faoláin clearly approves 'Joyce's olympian indifference to the charms of Cathleen Ni Houlihan' (1969: 140). On the effect of Joyce's antipathy to realism see Eagleton, 1994, where he writes: 'Joyce's writing is the non-Irish-speaking Irish writer's way of being unintelligible to the British. By subverting the very forms of their language, he struck a blow for all his gagged and humiliated fellow country people. And if he could do this so finely, it was because he was heir to a set of cultural lineages for which realism, as the fruit of a developed colonialist civilisation, had never been anything but profoundly problematic' (26).
36. In the chapter entitled 'The Double Session' in his book *Dissemination* (1981: 201–55) Derrida notes a similar operation occurring in a poem by Mallarmé in which notions of the truth

of literary representation are constantly played off against, shown to be the product of, the moment of representation. In noting correspondences such as this, we should also acknowledge that much of the theoretical revolution which has overtaken Western intellectual discourse in the last thirty years or so is influenced, through one line or another, by Joyce; and many of the issues which exercise the contemporary intellectual imagination – language, gender, exile, and so on – are anticipated in his writings.

37. McGreevy's essay, 'The Catholic Element in Work in Progress' (in Beckett, 1961: 119–27), located Joyce's work in a tradition of European Manichaen theology which, drawing on the philosophy of Aquinas, preached the co-existence of good and evil. By pointing to a long history of the exploration of the nature of evil by Catholic scholars, McGreevy highlighted the moral squeamishness and intellectual patronism underpinning contemporary Irish condemnations of Joyce's work.
38. I have argued elsewhere (1996b) that although Joyce's work exposes criticism's traditional pretensions to master the artistic text from some secure extra-textual location, it is difficult to imagine how such a discourse might be harnessed to forms capable of practical critical intervention. This of course is precisely the point, but it does leave potentially disruptive decolonising practices open to reappropriation by the modes of liberal and radical decolonisation against which Joyce sets his artistic stall. Criticism's inevitable dedication to the revelation of meaning, its 'will-to-reference', militates against the moment of critical erasure, no matter how understanding the critic or how disruptive the text.

Chapter 4

1. See Anne McClintock, 'The Angel of Progress: Pitfalls of the Term "Post-colonialism"' in Barker, Hulme and Iverson, 1994: 253–66.
2. Brown, 1985: 211–38; Lee, 1989: 271–328; Lyons, 1973: 559–628. Even Luke Gibbons, who is otherwise concerned to expose the periodisation of modern Irish history and what he calls 'the myth of modernization in Irish culture' (1996: 82–94), concedes that the 1950s was a decade of depression. Despite my intention to reveal the persistence of (de)colonialist modes of thought well into the 'postcolonial' period, I would

nevertheless concur with his assertion that 'The bogey of traditionalism and rural values can no longer be used in Ireland as a scapegoat for a regressive politics that emanates from the metropolitan centre' (92–3). On the contemporary ignorance of Irish intellectual culture during the 1950s see McCormack, 1995.

3. Brown, 1985: 236–38. In *Dead as Doornails* (1976) and *Remembering How We Stood* (1975), Anthony Cronin and John Ryan depict the grim social and artistic milieu of mid-century Dublin in which writers such as Kavanagh, Behan and O'Brien were trying to work.
4. For a contemporary discussion of these issues see Donoghue, 1955; Davie, 1955; and Mercier, 1956. Consistently the most perspicacious and self-conscious Irish literary intellectual of the period, Mercier invited his interlocutors to address what he perceived to be the dangerous absence of 'an Irish school of criticism' (87). He identified three main manifestations of this absence: the lack of a literary history of Anglo-Irish literature; the paucity of bibliographical and archival research; the gap between professional academic scholars and amateur literary critics.

Chapter 5

1. In his essay 'Between Politics and Literature: The Irish Cultural Journal' (1988: 250–68), Richard Kearney locates the roots of this split in the dualistic nature of the journal form itself, on the one hand towards subjective vision and on the other towards objective accuracy.
2. Although Terence Brown has suggested to me that its lack of influence diminishes any theory based upon an analysis of *Kavanagh's Weekly*, it is the very marginality of this journal when compared with a (relatively) successful publication such as *The Leader* (an important source for Brown's *Ireland: A Social and Cultural History 1922–1985*) that I find interesting here. I have no doubt as to the importance of *The Leader* in Irish cultural politics of the period, or, in the words of Joe Lee, that it 'reached an exceptional level of intellectual sophistication on both national and international affairs' (1989: 606). But the economy of 'importance' and 'marginality' is precisely the issue; the question of 'major' and 'minor' criticisms engages

with the same kinds of decolonising issues explored by David Lloyd in his analysis of 'major' and 'minor' literatures (1987).

3. *Kavanagh's Weekly: A Journal of Literature and Politics* vol.1, nos 1–13; Dublin: 12 April–5 July 1952. All future journal references in this chapter will cite volume and date (where relevant), number and page number – e.g., *Kavanagh's Weekly* number 3, page 5, will appear as 3:5.
4. From 'The Bellman (Larry Morrow): Meet Patrick Kavanagh' (18 April 1948), republished in Kavanagh, 1987: 117–22 (120).
5. Kavanagh already had experience of Irish libel law, having being involved in a suit brought against his book *The Green Fool* by Oliver St John Gogarty in 1939. Shortly after the demise of *Kavanagh's Weekly* he was to experience it again, this time as plaintiff against what he considered a derogatory article published in *The Leader* in October, 1952.
6. In 'The Social Role of the Literary Critic' Judith R. Kramer writes: 'The established critics of the twentieth century in both England and America are virtually all faculty members of universities. Their primary source of income is their academic salaries, although many of them are also editors of the critical quarterlies published at the universities. As a faculty member of a university, the critic is no longer a "free lance" intellectual, but rather, a salaried employee of an academic bureaucracy ... In essence, this amounts to a new form of "institutional" patronage' (in Albrecht, Barnett and Griff, 1970: 442).
7. Quinn writes: 'Prose and poetry mingled casually and unselfconsciously in this urban journalism. Not only did several of Kavanagh's poems make their first appearance in an opinion column; on occasion, they substituted for an article or book review. The demarcation between his creative and his critical writings diminished as the same preoccupations and attitudes, the same words, and above all, the same speaking voice were encountered in both' (1991: 254–55). See also Heaney's two essays – 'From Monaghan to the Grand Canal: The Poetry of Patrick Kavanagh' (1980: 115–30) and 'The Sense of Place' (1980: 131–49) – for an analysis of Kavanagh's poetic style which, it is claimed, 'has a spoken rather than a written note' (116).
8. As part of his contribution to the *Envoy* special number dedicated to Joyce, Kavanagh submitted the poem 'Who Killed James Joyce', containing the lines: 'Who killed James Joyce? I, said the commentator, I killed James Joyce For my

graduation. What weapon was used To slay the mighty Ulysses? The weapon that was used Was a Harvard thesis. How did you bury Joyce? In a broadcast Symposium. That's how we buried Joyce To a tuneful encomium. And did you get high marks, The Ph.D? I got the B.Litt And my master's degree.'

9. In his essay 'Swift's Tory Anarchy' (1991: 54–71), Said develops a similar argument with regard to the work of Jonathan Swift in the early eighteenth century.
10. I have been unable to discover circulation figures for *Kavanagh's Weekly*, although its lack of sponsorship and Anthony Cronin's remarks (1976) on the number of unsold copies which ended up as firing in the Pembroke Road flat indicate the journal's lack of commercial success.
11. In 'A Note on the Journal Genre' (1988: 250–68) Richard Kearney writes: 'Its aim was to invite the reader to participate in the multiple interpretations of events represented by the different essays. Indeed the very term "essay" is perhaps significant here, for it acknowledges that the various contributions printed in a single issue of a journal are no more than a variety of *attempts* (French, *essais*) to reach a consensus. The essays of a magazine are parts in search of a whole, diverse perspectives which require the reflective response of the reader if they are to achieve any sort of overall synthesis' (268). *Kavanagh's Weekly* lacks this sense of diversity, sacrificing diversity for a consistent confrontational tone which demands only assent or dissent, but never synthesis.
12. Many subsequent commentators have nominated O'Faoláin as the heir to AE's mantle of Irish writer as social critic. See for example Deane, 1986; Harmon, 1984; O'Dowd, 1985, 1991.
13. *Kavanagh's Weekly* was eight double-columned pages long and sold for sixpence; *The Bell* averaged about sixty pages and cost one and a half shillings. Whereas the former obtained only two regular commercial advertisements, the latter had a constant supply and turnover of commercial support.
14. Much of O'Faoláin's output during this period – in texts such as *The Irish* (1947), 'The Dilemma of Irish Letters' (1949), 'On Being an Irish Writer' (1953), etc. – concerns the relationship between literature, criticism and nation, and affects a holistic perspective with regard to Irish experience. See the 'Bibliography' of his work contained in Harmon, 1984: 214–32. On other contemporary Irish reactions to American 'New Criticism' see Davie, 1955; Donoghue, 1955; Mercier 1956.

15. In 'Hopes and Fears for Irish Literature' (first published in the journal *United Ireland*, 15 October 1892), comparing the 'mere verbal beauty of the newest generation of literary men in France and England' with Ireland, where he considered 'the art of living interests us greatly, and the art of writing but little', Yeats stated: 'Can we but learn a little of their skill, and a little of their devotion to form, a little of their hatred of the commonplace and the banal, we may make all these restless energies of ours alike the inspiration and the theme of a new and wonderful literature' (Storey, 1988: 74–7; 76).
16. '*The Bell* is quite clear about certain practical things and will, from time to time, deal with them – the Language, Partition, Education, and so forth ... For we eschew abstractions and will have nothing to do with generalisations that are not capable of proof by concrete experience' (O'Faoláin, 1:1:1).
17. David Marcus, 'From Joyce to Joyce', 18 (April 1952): 43.
18. It was Daniel Corkery who claimed that 'normal and national are synonymous in literary criticism' (1931: 3).
19. Temple Lane, 'Self Interview' in *Poetry Ireland* (18:3–9, July 1952); Lord Dunsany, 'Book Review' in *Irish Writing* (19:53–4, June 1952); Valentin Iremonger, 'Dramatic Criticism in Dublin' in *The Bell* (8:3:184–8, June 1952).
20. One of the regulations of the School of Celtic Studies founded under the auspices of the government-funded Dublin Institute for Advanced Studies (see below) provided for a Statutory Public Lecture to be delivered each year. However, as many of these lectures appeared later as monographs, books and pamphlets evincing the same characteristics already noted, there is no reason to believe that this was an attempt at popularising the practice or harnessing it for any practical purposes.
21. Quoted in Myles Dillon, 'The Dublin School of Celtic Studies', in Sommerfelt, 1958: 264–7 (264).
22. Several modern critics and philosophers have commented on the technologisation of the modern world and the loss in political effectivity which invariably accompanies the rise in intellectual specialisation. See Foucault 1970, 1979, 1981; Habermas, 1979, 1984; O'Dowd, 1985, 1988; Said, 1991.
23. The first, written by Brian O'Cuiv, appeared in *Ériu*, 16:171–79, 1952; the second, by James Carney, in *Celtica*, 1:86–110, 1950.
24. From a review quoted in an advertisement in *Irish Historical Studies*, September 1949.

25. Donald A. Davie, 'On the Poetic Diction of John M. Synge' (32–8). Davie was a lecturer in Trinity College in the early 1950s and also an immensely prolific critic and poet. See the section on 'The University' in ch. 6 below.
26. As well as the titles already mentioned, the list of journals edited at Maynooth at this time included *Archivum*, *Irish Ecclesiastical Record*, *Irish Theological Quarterly* and *Silhouette*.
27. On the roots of Trinity's intellectual conservatism and its 'distinctively Irish-Protestant cast of mind' see Foster, 1989: 124ff.; McDowell and Webb, 1982. In his analysis of the experiences of Conor Cruise O'Brien and Vivian Mercier at Trinity during the 1940s, McCormack notes 'a still entrenched loyalism in the college' (1995: 81).
28. Coghill had served in the British Army in both world wars, being mentioned in despatches for bravery on the Western Front. A footnote reveals that 'Somerville and Ross' was first 'delivered as an address to the Friends of the Library in June 1951' (47).
29. On the role of humour and double-vision in Irish cultural debate see 'Beckett: The Demythologising Intellect' in Kearney, 1985: 267–93.

Chapter 6

1. Bennett, 1990: 242. 'My main point', Bennett continues (and it is one with which I concur, and which this chapter is designed to address), 'is that generalised conceptions of criticism's function, especially when they rest on lapsarian accounts of criticism's history, deny any space within which questions of such a detailed and specific kind can even be put' (242).
2. Baldick describes the 'judicial' as one of literary criticism's two major constitutive metaphors (the other is the economic): 'In casting themselves as "judges" or as witnesses for the defence, critics habitually mimic the authority of more powerful assessors of literature ... Criticism from Plato onwards has ... presupposed censorship, banishment, and official persecution in the very language of its "judgements" and in its images of its own authority' (1983: 8–9).
3. Q.D. Leavis's *Fiction and the Reading Public* (1932) and Julien Benda's *La Trahison des clercs* (1928) are examples of contemporary texts which are concerned with the intellectual's response to modern mass culture.

4. This point was astutely observed by Austin Clarke during a period of heightened activity by the Censorship Board. In an article on the censorship in the *New Statesman* entitled 'Banned Books' he wrote: 'The indirect value of culture as a commercial asset has been recognised in many countries since the last war, and in the Irish Republic the fact has not escaped notice. The Irish contribution to modern literature has been a remarkable one, and the temptation to exploit it now has proved irresistible. Old quarrels have been forgotten and the names of W.B. Yeats, Joyce, Synge and others have become useful' (1953: 606).
5. See the correspondence between Patrick Kavanagh, Robert Greacen and Ewart Milne in the *New Statesman*, February 1949, for some of the standard contemporary arguments.
6. 'No Catholic may enter the Protestant University of Trinity College without having previously submitted his case to the Ordinary of the Diocese. Any Catholic who deliberately disobeys this law is guilty of a mortal sin, and while he persists in disobedience is unworthy to receive the Sacraments.' Quoted in Inglis, 1952a: 289. The 'ban' was finally removed in 1970.
7. Bennett argues 'that the moment of criticism's academic institutionalisation, particularly when viewed in the light of its subsequent extension throughout the education system, enormously augmented its power as an effective social force' (1990: 238). He goes on to call 'for a history of criticism that is less concerned with the move from one significant critic or school of criticism to another than with the development of institutionally embedded forms of instruction, training and examination. Not all practices of textual commentary acquire their social effectivity by organising the reader as a subject who takes a meaning from the text, with subsequent consequences for her/his consciousness and mode of relating to and acting within a generalised public arena. Others do so by producing the reader as an agent who performs a practice within specific institutional domains to become the bearer of specific certified competences. Some fulfil these two functions simultaneously' (239). Bennett's position owes something to Louis Althusser who argued 'that the ideological State apparatus which has been installed in the *dominant* position in mature capitalist formations as a result of a violent political and ideological class struggle against the old dominant ideological State apparatus, is the *educational ideological apparatus*' (1971: 152, original emphases). In fact, Althusser's thoughts (albeit class-based) on the relations

between censorship, education and ideology are highly suggestive for what follows.

8. These figures and percentages, and the ones which follow, have been abstracted from information included in the annual TCD calendars of this period.
9. Donald Davie, 'The Poetic Diction of J.M. Synge', *The Dublin Magazine* (January–March 1952), 32–8. Davie's academic and poetic interests arise specifically from the contemporaneous British debates surrounding the 'Movement'. For an analysis of his article on Synge see ch. 5 above.
10. See his articles: '*Ulysses* from Homer to Joyce. A Paper Delivered to the Classical Association at Manchester' (1949); 'The Mysticism that Pleased Him: a Note on the Primary Source of Joyce's *Ulysses*' in *Envoy*, 17 (1951), 62–9; 'Joyce's First Meeting with *Ulysses*', *The Listener* 1168 (1951), 99, 105; 'Ulyssean Qualities in Joyce's Leopold Bloom', *Comparative Literature* 5 (1953), 125–36, materials subsequently utilised in his *The Ulysses Theme* (1954).
11. In Table 2 the Social Science category includes subjects such as Psychology, Anthropology, Ethnography, Politics, Sociology, Economics, Criminology, Law, Education, Geography and History. Science includes Mathematics, Astronomy, Astrophysics, Physics, Chemistry, Geology, Biology, Medicine, Technology and Engineering. The Humanities category includes Philosophy, Religion, Art, Music, and History and Philosophy of Science. Literary Criticism covers material in the area of English, Old English, Irish, French, German, Spanish, Greek, Latin and Semitic.
12. Liam O'Dowd defines Unionism as 'a movement in which political intellectuals have been traditionally marginalised and constrained by popular political culture' (1991: 153).
13. Although there is no space here for an analysis of such texts, even their titles and the critical assumptions which appear to underpin them make for interesting reading in the light of present arguments.
14. O'Dowd writes that 'All the major analyses of modern nationalist movements have detailed the crucial role of intellectuals, especially those of a literary and humanistic orientation. This role has been more autonomous in regions with a similar class structure to nationalist Ireland, with large farming populations, a dispersed urban working class with limited industrial experience, and a substantial commercial and small business class' (1991: 161).

15. Bernadette Whelan has argued that it was not until after the period under examination here – that is, 1958 onwards – that research culture began to pick up in the Republic, and even then, it is clear that 'arts/humanities subjects' suffered in comparison with 'business, pure science, and the social science subjects' (1996: 173). Whelan quotes the Commission on Higher Education's *Presentation and Summary of Report* (Dublin: Government Publications Office, 1967: 73) that 'Postgraduate and research work is a prime factor of the university. But in this country it has remained comparatively under-developed ... The provision of a greatly increased number of adequate postgraduate scholarship and fellowships is a necessity' (175). Such an initiative, however, imagines a specific role for the university and the academic critic as both are enlisted to play a part 'in national policies of research directed toward the country's *economic* welfare' (1967: 80, emphasis added). Moreover, as Whelan goes on to suggest, 'as research opportunities remained so narrowly focused on the American experience, it simply substituted one form of cultural isolationism for another' (175).
16. Major new editions of all the above-named authors, plus Swift (Davis, 1948, 1951, 1955), Wilde (Holland, 1948) and O'Flaherty (1956) were issued by publishers such as Blackwell, Macmillan and many of the American university presses, while for contemporary Irish writers a significant development was the founding of the Irish Book Club by the New York publishers Devin Adair Garrity in 1948.
17. Editing *An Anthology of Irish Literature* (1954), Greene argued a similar line when he wrote that: 'The fact is that the literature of the last two centuries and a half is bilingual, and one is forced to describe all of it as Irish' (xxxii). Mercier, for his part, believed that the task of 'conscientious' popular anthologising – crucial for any 'Irish school of criticism' worthy of the name – was being ignored by the state's professional scholars (1956: 87, 86).
18. In 1948, out of 615 publishers listed in *The British* [sic] *Writers' and Artists' Year Book* only fifteen were Irish. These included Browne & Nolan, The Catholic Truth Society of Ireland, Clonmore & Reynolds, Duffy & Co., Eason & Son, Gill & Son, Hodges, Figgis & Co., The Mercier Press, The National Press, The Talbot Press and The Quota Press. By 1958 this number had dropped to fourteen.

Chapter 7

1. Although the rise of Irish-related professional academic discourse was part of a wider Western phenomenon, there are determinants, forms and narratives peculiar to a decolonising society, and these shall be the focus of this chapter.
2. Some of the more significant moments in this process include A. Norman Jeffares, *W.B. Yeats: Man and Poet* (1949) and *The Poetry of W.B. Yeats* (1961); Hugh Kenner, *Dublin's Joyce* (1955) and *Samuel Beckett: A Critical Study* (1961); Frank Kermode, *Romantic Image* (1957); David Krause, *Sean O'Casey: The Man and His Work* (1960); Richard Kain, *Fabulous Voyager: James Joyce's Ulysses* (1947); William York Tindall, *James Joyce* (1950); F.A.C. Wilson, *W.B. Yeats and Tradition* (1958).
3. George Watson disputes this description of the critical monograph, arguing 'that the natural unit of descriptive criticism ever since Dryden has been the essay or article, a book hundreds of pages long being an unnatural phenomenon at best' (1964: 217). See the section on 'Truth and Method' below.
4. Quoted in *Kavanagh's Weekly* (2:3) Saturday, 19 April 1952.
5. When Gerard Murphy's *Saga and Myth in Ancient Ireland* was republished in 1971, the list of series titles at the back read as follows: I. *Theatre in Ireland* by Micheál MacLiammóir; II. *Poetry in Modern Ireland* by Austin Clarke; III. *Irish Folk Music and Song* by Donal O'Sullivan; IV. *Irish Landscape* by R. Lloyd Praeger; V. *Conamara* by Seán Mac Giollarnáth; VI. *Irish Classical Poetry* by Eleanor Knott. VII. *The Personality of Leinster* by Maurice Craig; VIII. *Early Irish Society* edited by Myles Dillon; IX. *The Fortunes of the Irish Language* by Daniel Corkery; X. *Saga and Myth in Ancient Ireland* by Gerard Murphy; XI. *The Ossianic Lore and Romantic Tales of Medieval Ireland* by Gerard Murphy; XII. *Social Life in Ireland 1800–45* edited by R.B. McDowell; XIII. *Dublin* by Desmond F. Moore; XIV. *The Irish Language* by David Greene; XV. *Irish Folk Custom and Belief* by Seán Ó Súilleabháin; XVI. *The Irish Harp* by Joan Rimmer; XVII. *Eriugena* by John J. O'Meara.
6. Such an intervention recalls developments in England half a century earlier when, as Brian Doyle puts it (in 'The Invention of English' in Colls and Dodd, 1986: 89–115), 'A number of educationalists, politicians, philosophers and political theorists searched for new and more efficient ways of building and disseminating a national sense of ancestry, tradition and universal "free" citizenship' (89), and found that culture, and especially

English literature, might function as a novel form 'to resolve a number of problems for the functioning of national institutions between 1880 and 1920' (110–11).

7. The notion of 'passing' (in the sense of passing oneself off as something one is not) is derived in the first instance from Nella Larsen and her novel of that title (1929), where it is described as 'this breaking away from all that was familiar and friendly to take one's chance in another environment' (1989: 263). Conceived thus, 'passing' offers a variation on the difficult and dangerous freedoms postulated by Bhabha and Spivak. However, there has as yet been no systematic or comprehensive theoretical analysis of the dynamics of 'passing' within a racial, gender or colonial framework.
8. This was an argument which Seamus Heaney (in many of the essays collected in *Preoccupations*, 1980) was to find attractive, and which is still influential in some critical circles.
9. In a review of the book published in *The Spectator* ('Irish Background', 189:226, 15 August 1952) Patrick Kavanagh attacked Ussher, first of all, for knowing nothing about the 'Irish', and secondly, for trying to claim authority on the grounds of his nationality. This contradictory critique is typical of Kavanagh, as he simultaneously accepts and denies national experience, in much the same way as his target.
10. Ussher's genealogy of the trouble-making, double-thinking 'Irish mind' anticipates Richard Kearney's similar project (1985) by some three and a half decades, and is subject to the same advantages and drawbacks in relation to received decolonising discourses. See the section 'Decolonisation in Ireland' in ch. 1.

References

All titles published in London unless otherwise stated.

Abrams, M.H. (Gen. Ed.), *The Norton Anthology of English Literature,* vol.1 (4th edn, New York: Norton & Co., 1979).

Adams, Michael, *Censorship: The Irish Experience* (Dublin: Scepter Books, 1968).

Adorno, Theodor W. and Max Horkheimer, *Dialectic of Enlightenment* [1947], trans. J. Cumming (Verso, 1979).

Albrecht, M.C., J.H. Barnett and M. Griff (eds), *The Sociology of Art and Literature: A Reader* (Duckworth, 1970).

Althusser, Louis, *Lenin and Philosophy and Other Essays*, trans. B. Brewster (New York: Monthly Review Press, 1971).

Anderson, Benedict, *Imagined Communities: Reflections on the Origin and Spread of Nationalism* [1983] (rev. edn, Verso, 1991).

Anonymous (ed.), *The Transactions of the Royal Irish Academy MDCCLXXXVII* (Dublin: George Bonham, for the Academy, 1787).

Anonymous (ed.), *Thomas Davis 1845–1945: Essays and Poems with a Centenary Memoir by Eamon de Valera* (Dublin: Gill and Son Ltd, 1945).

Arnold, Bruce, *The Scandal of Ulysses* (Sinclair-Stevenson, 1991).

Arnold, Matthew, *Selected Poetry and Prose*, ed. F.L. Mulhauser (New York: Holt, Rinehart and Winston, 1953).

Ashcroft, Bill, Gareth Griffiths and Helen Tiffin (eds), *The Post-Colonial Studies Reader* (Routledge, 1995).

Atkins, J.W.H., *English Literary Criticism: 17th and 18th Centuries* (Methuen, 1951).

Attridge, Derek (ed.), *Jacques Derrida: Acts of Literature* (Routledge, 1992).

Attridge, Derek and Daniel Ferrer (eds), *Post-structuralist Joyce: Essays from the French* (Cambridge: Cambridge University Press, 1984).

Baldick, Chris, *The Social Mission of English Criticism 1848–1932* (Oxford: Clarendon Press, 1983).

Barker, Francis, Peter Hulme and Margaret Iversen (eds), *Colonial Discourse/Postcolonial Theory* (Manchester: Manchester University Press, 1994).

Barnett, Ronald, *Realizing the University* (Institute of Education University of London, 1997).
Barry, Kevin, 'Critical Notes on Post-Colonial Aesthetics', *Irish Studies Review* 14 (Spring 1996), 2–11.
Barthes, Roland, 'Criticism as Language' in Lodge, 1972: 647–51.
Beckett, Samuel, *et al.*, *Our Exagmination Round His Factification For Incamination Of Work In Progress* [1929] (Faber & Faber, 1961).
Belsey, Catherine, 'Literature, History, Politics' in Lodge, 1988: 400–10.
Benda, Julien, *The Treason of the Intellectuals* [1928], trans. R. Aldington (New York: Norton, 1969).
Bennett, Tony, *Outside Literature* (Routledge, 1990).
Bhabha, Homi K., 'Representation and the Colonial Text: A Critical Exploration of Some Forms of Mimeticism' in Frank Gloversmith (ed.), *The Theory of Reading* (Sussex: Harvester Press, 1984), 93–122.
—— 'Foreword: Remembering Fanon' in Fanon (1986), vii–xxv.
—— (ed.), *Nation and Narration* (Routledge, 1990).
—— *The Location of Culture* (Routledge, 1994).
Blanchard, Paul, *The Right to Read: The Battle Against Censorship* (Boston: Beacon Press, 1955).
Bodkin, Thomas, *Report on the Arts in Ireland* (Dublin: The Stationery Office, 1949).
Boyce, George D., *Nationalism in Ireland* (Beckenham: Croom Helm, 1982).
Brady, Anne M. and Brian Cleeve, *A Biographical Dictionary of Irish Writers* (Mullingar: Lilliput Press, 1985).
Brady, Ciarán (ed.), *Interpreting Irish History: The Debate on Historical Revisionism* (Dublin: Irish Academic Press, 1994).
Breathnach, R.A., 'Two Eighteenth-Century Irish Scholars: J.C. Walker and Charlotte Brooke', *Studia Hibernica* 5 (1965), 88–97.
Brown, Malcolm, *The Politics of Irish Literature: From Thomas Davis to W.B. Yeats* (George Allen & Unwin Ltd, 1972).
Brown, Terence, *Ireland: A Social and Cultural History 1922–1985* (Fontana, 1985).
—— Review of Ó Drisceoil (1996) in *Irish Studies Review* 20 (Autumn 1997), 45–6.
Burke, Seán, *The Death and Return of the Author: Criticism and Subjectivity in Barthes, Foucault and Derrida* (Edinburgh: Edinburgh University Press, 1992).
Burney, Charles, 'Review of Walker's *Irish Bards*', *Monthly Review* 77 (December 1787), 425–39.

Cairns, David and Shaun Richards, *Writing Ireland: Colonialism, Nationalism and Culture* (Manchester: Manchester University Press, 1988a).

—— 'Discourses of Opposition and Resistance in Late Nineteenth and Early Twentieth Century Ireland', *Text and Context* 2.1 (Spring 1988b), 76–84.

Carney, James, *Poems on the O'Reillys* (Dublin: Dublin Institute for Advanced Studies, 1950).

—— *Studies in Irish Literature in History* (Dublin: Dublin Institute for Advanced Studies, 1955).

Chambers, Iain and Lidia Curti (eds), *The Post-colonial Question: Common Skies, Divided Horizons* (Routledge, 1996).

Chatterjee, Partha, *Nationalist Thought and the Colonial World: A Derivative Discourse?* (Zed Books, 1986).

Clarke, Austin, *Poetry in Modern Ireland* [1951] (Dublin: At The Sign of the Three Candles, 1961).

—— 'Banned Books', *New Statesman* 45 (Saturday, 23 May 1953), 606.

Clifford, J.L. (ed.), *Man versus Society in 18th-Century Britain: Six Points of View* (Cambridge: Cambridge University Press, 1968).

Cobban, Alfred, *Edmund Burke and the Revolt against the Eighteenth Century: A Study of the Political and Social Thinking of Burke, Wordsworth, Coleridge and Southey* [1929] (George Allen & Unwin Ltd, 1960).

Colls, Robert and Philip Dodd (eds), *Englishness: Politics and Culture 1880–1920* (Beckenham: Croom Helm, 1986).

Con Davis, Robert and Ronald Schleifer, *Criticism and Culture: The Role of Critique in Modern Literary Studies* (Longman, 1991).

Connolly, James, *Labour in Irish History* [1910] (Dublin: New Books Publications, 1983).

Corkery, Daniel, *The Hidden Ireland: A Study of Gaelic Munster in the Eighteenth Century* [1924] (Dublin: Gill and Son, 1967).

—— *Synge and Anglo-Irish Literature: A Study* (Cork: Cork University Press, 1931).

—— *The Fortunes of the Irish Language* (Dublin: At The Sign of the Three Candles, 1954).

Cronin, Anthony, *Dead as Doornails: A Chronicle of Life* (Dublin: Dolmen Press, 1976).

Curtis, Edmund, *A History of Ireland* [1936] (Methuen, 1961).

Davie, Donald, 'Reflections of an English Writer in Ireland', *Studies: An Irish Quarterly Review of Letters, Philosophy and Science* 44 (Winter 1955), 439–45.

Davis, Herbert (ed.), *The Prose Writings of Jonathan Swift* (IX, XI, XII) (Oxford: Blackwell, 1948, 1951, 1955).
Deane, Seamus, 'Heroic Styles: The Tradition of an Idea' in Field Day Theatre Company (1985), 45–58.
—— *A Short History of Irish Literature* (Hutchinson 1986).
—— (ed.) *Nationalism, Colonialism and Literature* (Minneapolis: University of Minnesota Press, 1990).
—— (Gen. Ed.), *The Field Day Anthology of Irish Writing,* 3 vols (Derry and London: Field Day and Faber, 1991).
Deleuze, Gilles and Félix Guattari, *Anti-Oedipus: Capitalism and Schizophrenia*, vol. I [1972], trans. R. Hurley, M. Sheen and H. Lane (New York: Viking, 1977).
De Man, Paul, *Blindness and Insight: Essays in the Rhetoric of Contemporary Criticism* [1971] (Methuen, 1983).
Deming, R.H. (ed.), *James Joyce: The Critical Heritage*, vols I (1907–1927) and II (1928–1941) (Routledge & Kegan Paul, 1970a and 1970b).
Derrida, Jacques, *Of Grammatology* [1967], trans. G.C. Spivak (Johns Hopkins University Press, 1976).
—— *Dissemination* [1972], trans. B. Johnson (Chicago: University of Chicago Press 1981).
—— *Mémoires: for Paul de Man*, trans. C. Lindsay, J. Culler and E. Cadava (New York: Columbia University Press, 1986).
Dillon, Myles, *Early Irish Literature* (Chicago: University of Chicago Press, 1948).
—— (ed.) *Irish Sagas* (Dublin: The Stationery Office, 1959).
Docherty, Thomas, 'Tragedy and the Nationalist Condition of Criticism', *Textual Practice* 10.3 (Autumn 1996), 479–505.
Donoghue, Denis, 'Notes Towards a Critical Method', *Studies: An Irish Quarterly Review of Letters, Philosophy and Science* 42 (Summer 1955), 185–92.
Duffy, Charles Gavan (ed.), *The Revival of Irish Literature and Other Addresses* (Fisher Unwin, 1894).
Eagleton, Terry, *Criticism and Ideology: A Study in Marxist Literary Theory* [1976] (Verso, 1978).
—— *Literary Theory: An Introduction* (Oxford: Basil Blackwell, 1983).
—— *The Function of Criticism: From The Spectator to Post-Structuralism* (Verso, 1984).
—— *Saint Oscar* (Derry: Field Day, 1989).
—— 'Form and Ideology in the Anglo-Irish Novel', *Bullán: An Irish Studies Journal* 1.1 (Spring 1994), 17–26.

—— *Heathcliff and the Great Hunger: Studies in Irish Culture* (Verso, 1995).

Eco, Umberto, *The Middle Ages of James Joyce* [1962] Hutchinson Radius, 1982).

Eglinton, John, *Irish Literary Portraits* (Macmillan & Co. Ltd, 1935).

Ellmann, Richard, *Yeats: The Man and the Mask* [1948] (Faber, 1961).

—— *The Identity of Yeats* (London and New York: Macmillan, 1954).

—— *James Joyce* (New York: Oxford University Press, 1959).

Fanon, Frantz, *Black Skin, White Masks* [1952], trans. C.L. Markmann (Pluto Press, 1986).

—— *The Wretched of the Earth* [1961], trans. C. Farrington (Middlesex: Penguin, 1967).

Farren, Robert, *The Course of Irish Verse in English* (Sheed and Ward, 1948).

Ferguson, Samuel, 'Hardiman's Irish Minstrelsy', *Dublin University Magazine*: Part One, 3.16 (April 1834), 465–77; Part Two, 4.20 (August 1834), 152–67; Part Three, 4.22 (October 1834), 447–67; Part Four, 4.23 (November 1834), 514–42.

Field Day Theatre Company, *Ireland's Field Day* (Hutchinson, 1985).

Flood, W.H.G., *A History of Irish Music* (Dublin: Browne and Nolan Limited, 1905).

Flower, Robin, *The Irish Tradition* (Oxford: At The Clarendon Press, 1947).

Foster, Roy, 'We Are All Revisionists Now', *Irish Review* 1 (1986), 1–5.

—— *Modern Ireland 1600–1972* [1988] (Penguin, 1989).

—— *Paddy and Mr Punch: Connections in Irish and British History* (Allen Lane, 1993).

Foucault, Michel, *The Order of Things: An Archaeology of the Human Sciences* [1966] (Tavistock Publications Limited, 1970).

—— 'The Order of Discourse' [1970] in Young, 1981: 48–78.

Gellner, Ernest, *Nations and Nationalism* (Oxford: Basil Blackwell, 1983).

—— *The History of Sexuality, Vol. I: An Introduction*, trans. R. Hurley (Allen Lane, 1979).

Genette, Gérard, 'Structuralism and Literary Theory' in Lodge, 1988: 63–78.

Gibbons, Luke, 'Challenging the Canon: Revisionism and Cultural Criticism' in Deane 1991, vol. III, 561–8.

—— 'Dialogue Without the Other? A Reply to Francis Mulhern', *Radical Philosophy: A Journal of Socialist and Feminist Philosophy* 67 (Summer 1994), 28–31.

—— *Transformations in Irish Culture* (Cork: Cork University Press, 1996).

Giddens, Anthony, *The Nation-State and Violence: Part Two of a Contemporary History of Historical Materialism* (Berkeley: University of California Press, 1987).

Graham, Colin, 'History, Gender and the Colonial Moment: *Castle Rackrent*', *Irish Studies Review* 14 (Spring 1996), 21–4.

Gramsci, Antonio, *Selections from the Prison Notebooks*, (ed. and trans. Q. Hoare and G. Nowell Smith (Lawrence and Wishart, 1971).

Greacen, Robert, 'Reply to Patrick Kavanagh', *New Statesman* XXXVII.936 (12 February 1949), 154.

Greacen, Robert and Valentin Iremonger (eds), *Contemporary Irish Poetry* (Faber & Faber Limited, 1949).

Greene, David (ed.), *An Anthology of Irish Literature* (New York: New York University Press, 1954).

Habermas, Jürgen, *Communication and the Evolution of Society*, trans. T. McCarthy (Heinemann, 1979).

—— *Theory of Communicative Action, Volume 1: Reason and the Rationalization of Society* [1981], trans. T. McCarthy (Heinemann, 1984).

Harmon, Maurice, *Seán O'Faoláin: A Critical Introduction* (Dublin: Wolfhound Press, 1984).

Harvey, David, *The Condition of Postmodernity: An Enquiry into the Origins of Cultural Change* (Oxford: Basil Blackwell, 1989).

Hawthorn, Jeremy, *Unlocking the Text: Fundamental Issues in Literary Theory* (Edward Arnold, 1987).

Heaney, Seamus, *Preoccupations: Selected Prose 1968–1978* (Faber & Faber, 1980).

—— 'William Butler Yeats (1865–1939)' in Deane 1991, vol. II, 783–90.

Henn, T.R., *The Lonely Tower: Studies in the Poetry of W.B. Yeats* (Methuen, 1950).

Hill, Jacqueline, R., 'The Intelligentsia and Irish Nationalism in the 1840s', *Studia Hibernica* 2 (1980), 73–109.

Hoagland, Katherine (ed.), *1000 Years of Irish Poetry: The Gaelic and Anglo-Irish Poets from Pagan Times to the Present* (New York: Devin-Adair Company, 1947).

Hobsbawm, Eric and Terence Ranger (eds), *The Invention of Tradition* (Cambridge: Cambridge University Press, 1983).

Hoffman, F.J., C. Allen and C.F. Ulrich (eds), *The Little Magazine: A History and a Bibliography* (New Jersey: Princeton University Press, 1947).
Hohendahl, Peter Uwe, *The Institution of Criticism* (Cornell University Press, 1982).
Holland, Vivian (ed.), *Complete Works of Oscar Wilde* (Collins, 1948).
Holzapfel, R. (ed.), *An Index of Contributors to The Bell* (Blackrock, 1970).
Hughes, Eamonn (ed.), *Culture and Politics in Northern Ireland* (Milton Keynes: Open University Press, 1991).
Humm, Peter, Paul Stigant and Peter Widdowson (eds), *Popular Fictions: Essays in Literature and History* (Methuen, 1986).
Hyde, Douglas, *A Literary History of Ireland From Earliest Times to the Present Day* [1899] (Ernest Benn Limited, 1967).
Inglis, Brian, 'Irish Double-Thought', *The Spectator* 188 (7 March 1952a), 289.
—— 'Smuggled Culture', *The Spectator* 188 (28 November 1952b), 726.
—— *The Freedom of the Press in Ireland 1784–1841* (Faber & Faber, 1954).
Irish PEN, *Concord of Harps: An Irish P.E.N. Anthology of Poetry* (Dublin: Talbot Press Limited, 1952).
Jameson, Fredric, *The Political Unconscious: Narrative as a Socially Symbolic Act* (Methuen, 1981).
Jeffares, A. Norman, *W.B. Yeats: Man and Poet* (New York: Barnes and Noble, 1949).
—— *The Poetry of W.B. Yeats* (New York: Barnes and Noble, 1961).
Johnson, Lionel, 'Poetry and Patriotism' in Storey, 1988: 93–106.
Joyce, James, *A Portrait of the Artist as a Young Man* [1916] (St Albans: Panther 1977).
—— *Ulysses* [1922], ed. Jeri Johnson (Oxford: Oxford University Press, 1993).
—— *Finnegans Wake* [1939] (Faber & Faber, 1964).
Kain, Richard, *Fabulous Voyager: James Joyce's Ulysses* (Chicago: University of Chicago Press, 1947).
Kavanagh, Patrick, 'Letter on Irish Censorship', *New Statesman* XXXVII.935 (5 February 1949), 130.
Kavanagh, Peter (ed.), *Patrick Kavanagh: Man and Poet* [1973] (Newbridge: Goldsmith Press, 1987).
Kearney, Richard (ed.), *The Irish Mind: Exploring Intellectual Traditions* (Dublin: Wolfhound Press, 1985).

—— *Transitions: Narratives in Modern Irish Culture* (Manchester: Manchester University Press, 1988).
—— *States of Mind: Dialogues with Contemporary Thinkers on the European Mind* (Manchester: Manchester University Press, 1995).
Kedourie, Elie, *Nationalism* (Hutchinson, 1960).
Kenner, Hugh, *Dublin's Joyce* (Chatto and Windus, 1955).
—— *Samuel Beckett: A Critical Study* (Boston: Beacon, 1961).
Kermode, Frank, *Romantic Image* (Routledge and Kegan Paul, 1957).
Kiberd, Declan, 'Anglo-Irish Attitudes' in Field Day Theatre Company, 1985: 83–105.
—— 'Contemporary Irish Poetry' in Deane 1990, vol. III: 1309–16.
—— *Inventing Ireland* (Jonathan Cape, 1995).
Kiely, Benedict, *Poor Scholar: A Study of the Works and Days of William Carleton (1794–1869)* [1947] (Dublin: Talbot Press, 1972).
—— *Modern Irish Fiction: A Critique* (Dublin: Golden Eagle Books Ltd, 1950).
Knott, Eleanor, *Irish Classical Poetry, Commonly Called Bardic Poetry* (Dublin: At The Sign of the Three Candles, 1957).
Krause, David, *Sean O'Casey: The Man and his Work* (Macmillan, 1960).
Kristeva, Julia, *Revolution in Poetic Language* [1974] (Guildford, Surrey: Columbia University Press, 1984).
—— *Desire in Language: A Semiotic Approach to Literature and Art* [1977] Oxford: Basil Blackwell, 1981).
Lacan, Jacques, *Ecrits: A Selection* [1966], trans. A. Sheridan (Tavistock Publications Ltd, 1977).
Larsen, Nella, *Quicksand* [1928] and *Passing* [1929], ed. in one volume by D.E. McDowell (London: Serpent's Tail, 1989).
Laurence, D.H. (ed.), *The Bodley Head Bernard Shaw Collected Plays with their Prefaces*, vol. II (The Bodley Head, 1971).
Lazarus, Neil, 'National Consciousness and the Specificity of (Post) Colonial Intellectualism' in Barker, Hulme and Iversen, 1994: 197–220.
Leavis, Q.D., *Fiction and the Reading Public* (Chatto and Windus, 1932).
Lee, J.J. (ed.), *Ireland 1945–70* (Dublin: Gill and Macmillan, 1979).
—— *Ireland 1912–1985: Politics and Society* (Cambridge: Cambridge University Press, 1989).
Leerssen, Joep, *Mere Irish and Fíor-Ghael: Studies in the Idea of Irish Nationality, its Development and Literary Expression Prior to the*

Nineteenth Century (2nd edn; Cork: Cork University Press in association with Field Day, 1996a).

—— *Remembrance and Imagination: Patterns in the Historical and Literary Representation of Ireland in the Nineteenth Century* (Cork: Cork University Press in association with Field Day, 1996b).

Lentricchia, Frank, *Criticism and Social Change* (Chicago: University of Chicago Press, 1983).

Lloyd, David, *Nationalism and Minor Literature: James Clarence Mangan and the Emergence of Irish Cultural Nationalism* (Berkeley: University of California Press, 1987).

—— *Anomalous States: Irish Writing and the Post-Colonial Moment* (Dublin: Lilliput Press, 1993).

—— 'Cultural Theory and Ireland: Review of Terry Eagleton's *Heathcliff and the Great Hunger: Studies in Irish Culture*', *Bullán: An Irish Studies Journal* 3.1 (Spring 1997), 87–91.

Lodge, David (ed.), *Twentieth-Century Literary Criticism* (Harlow, Essex: Longman, 1972).

—— *Modern Criticism and Theory: A Reader* (Longman, 1988).

Longley, Edna, *The Living Stream: Literature and Revisionism in Ireland* (Newcastle upon Tyne: Bloodaxe Books, 1994).

—— 'Irish Studies: A Forward Look', *British Association for Irish Studies Newsletter* 10 (Winter 1996, Spring 1997), 4–5.

Lyons, F.S.L., *Ireland Since the Famine* [1971] (rev. edn, Fontana/Collins, 1973).

Mac Aonghusa, Pronsias and Liam Ó Réagáin (eds), *The Best of Pearse* (Cork: Mercier Press, 1967).

MacCabe, Colin, *James Joyce and the Revolution of the Word* (Macmillan, 1979).

McCormack, W.J., *Ascendancy and Tradition in Anglo-Irish Literary History from 1789 to 1939* (Oxford: Clarendon Press, 1985).

—— *Dissolute Characters: Irish Literary History through Balzac, Sheridan le Fanu, Yeats and Bowen* (Manchester: Manchester University Press, 1993).

—— *From Burke to Beckett: Ascendancy, Tradition and Betrayal in Literary History* (Cork: Cork University Press, 1994).

—— 'Convergent Criticism: The *Biographia Literaria* of Vivian Mercier and the State of Irish Literary History', *Bullán: An Irish Studies Journal* 2.1 (Summer 1995), 79–100.

MacDonagh, Donagh and Lennox Robinson (eds), *The Oxford Book of Irish Verse: XVIIth Century–XXth Century* (Oxford: At The Clarendon Press, 1958).

MacDonagh, Thomas, *Literature in Ireland: Studies Irish and Anglo-Irish* (Dublin: Talbot Press Limited, 1916).

McDowell, R.B., *Ireland in the Age of Imperialism and Revolution 1760–1801* (Oxford: Clarendon Press, 1979).
McDowell, R.B. and D.A. Webb, *Trinity College, Dublin, 1592–1952: An Academic History* (Cambridge: Cambridge University Press, 1982).
MacLiammóir, Micheál, *Theatre in Ireland* (Dublin: At The Sign of the Three Candles, 1950).
Mannoni, D.O., *Prospero and Caliban: The Psychology of Colonization* [1956], trans. P. Powesland (New York: Praeger, 1964).
Memmi, Albert, *The Colonizer and the Colonized* [1957], trans. H. Greenfeld (Souvenir Press, 1974).
Mercier, Vivian, 'An Irish School of Criticism?', *Studies: An Irish Quarterly Review of Letters, Philosophy and Science* 45 (Spring 1956), 84–7.
Mercier, Vivian and David H. Greene (eds), *1000 Years of Irish Prose: Part 1 – The Literary Revival* [1952] (New York: Devin-Adair Company, 1953).
Milne, Ewart, 'Reply to Patrick Kavanagh', *New Statesman* XXXVII.935 (12 February 1949), 155.
Moi, Toril, *Sexual/Textual Politics* (Methuen & Co. Ltd, 1985).
Moody, T.W. and W.E. Vaughan, (eds), *A New History of Ireland: Volume IV – Eighteenth-Century Ireland 1691–1800* (Oxford: Clarendon Press, 1986).
Moore, George, *Impressions and Opinions* (T. Werner Laurie Limited, 1914).
—— *Avowals* [1919] (William Heinemann Limited, 1924).
Moran, David P., *The Philosophy of Irish Ireland* (Dublin: James Duffy & Co., 1905).
Mulhern, Francis, *The Moment of Scrutiny* [1979] (Verso, 1981).
—— 'A Nation, Yet Again: The Field Day Anthology', *Radical Philosophy: A Journal of Socialist and Feminist Philosophy* 65 (Autumn 1993), 22–9.
Murphy, Andrew, 'Reviewing the Paradigm: A New Look at Early-Modern Ireland', *Éire–Ireland: An Interdisciplinary Journal of Irish Studies* (Fall/Winter 1996), 13–40.
Murphy, Gerard, (ed.) *Early Irish Lyrics: Eighth to Twelfth Century* (Oxford: At The Clarendon Press, 1956).
Nairn, Tom, *The Break-Up of Britain: Crisis and Neo-Nationalism* (Verso, 1981).
Nandy, Ashis, *The Intimate Enemy: Loss and Recovery of Self under Colonialism* (Delhi: Oxford University Press, 1983).
Nolan, Emer, *James Joyce and Nationalism* (Routledge, 1994).
Norris, Christopher, *Derrida* (Fontana Press, 1987).

—— *Truth and the Ethics of Criticism* (Manchester: Manchester University Press, 1994).
O'Brien, Conor Cruise, *States of Ireland* (Hutchinson, 1972).
O'Connor, Frank (ed.), *Modern Irish Short Stories* (Oxford: Oxford University Press, 1957).
O'Dowd, Liam, 'Intellectuals in 20th Century Ireland: And the Case of George Russell (AE)', *The Crane Bag* 9.1 (1985), 6–25.
—— 'Neglecting the Material Dimension: Irish Intellectuals and the Problem of Identity', *Irish Review* 3 (1988), 8–17.
—— 'Intellectuals and Political Culture: A Unionist–Nationalist Comparison' in Hughes 1991: 151–73.
Ó Drisceoil, Donal, *Censorship in Ireland 1939–1945: Neutrality, Politics and Society* (Cork: Cork University Press, 1996).
O'Driscoll, Robert, *An Ascendancy of the Heart: Ferguson and the Beginnings of Modern Irish Literature in English* (Dublin: Dolmen Press, 1976).
O'Faoláin, Seán, *The Irish* [1947] (Penguin: 1980).
—— *The Short Story* (Collins, 1948).
—— 'The Dilemma of Irish Letters', *The Month* 2.6 (July–December 1949), 366–79.
—— 'On Being an Irish Writer', *The Spectator* (June 3 1953), 25–6.
—— *The Vanishing Hero: Studies in the Novelists of the Twenties* (Eyre and Spottiswoode, 1956).
O'Flaherty, Liam (ed.), *The Stories of Liam O'Flaherty* (New York: Devin-Adair, 1956).
Parry, Benita, 'Resistance Theory/Theorising Resistance, or Two Cheers for Nativism' in Barker, Hulme and Iversen, 1994: 172–96.
Passages, 'The Deconstruction of Actuality: An Interview with Jacques Derrida', *Radical Philosophy: A Journal of Socialist and Feminist Philosophy* 68 (Autumn 1994), 28–41.
Pêcheux, Michel, *Language, Semantics and Ideology: Stating the Obvious* [1975], trans. H. Nagpal (Macmillan, 1982).
Pinkney, Tony, 'To Criticise the Critic', *Literature and History* 8.2 (Autumn 1982), 248–53.
Plato, *The Republic* (no date, trans. A.D. Lindsay, 1935; J.M. Dent & Sons, 1976).
Price, Alan (ed.), *J.M. Synge: Collected Works – Volume 2: Prose* (Gerrards Cross, Bucks.: Colin Smythe, 1982).
Quinn, Antoinette, *Patrick Kavanagh: Born-Again Romantic* (Dublin: Gill and Macmillan, 1991).

Rafroidi, Patrick, *Irish Literature in English: The Romantic Period (1789–1850)* (Gerrards Cross, Bucks.: Colin Smythe, 1980).
Rogers, Pat (ed.), *The Context of English Literature: The Eighteenth Century* (Methuen & Co. Ltd, 1978).
Ryan, John, *Remembering How We Stood: Bohemian Dublin at Mid-Century* (Dublin: Gill and Macmillan Ltd, 1975).
Ryle, Martin, 'Long Live Literature? Englit, Radical Criticism and Cultural Studies', *Radical Philosophy: A Journal of Socialist and Feminist Philosophy* 67 (Summer 1994), 21–8.
Said, Edward W., *Orientalism* [1978] (Peregrine, 1985).
—— *The World, the Text, and the Critic* [1983] (Vintage, 1991).
—— *Culture and Imperialism* (Chatto & Windus, 1993).
Saintsbury, George, *A History of Criticism and Literary Taste in Europe, vols. I, II and III* (1900–4; 4th edn, Edinburgh and London: William Blackwood & Sons, 1949).
Salusinszky, Imre, *Criticism in Society: Interviews with Jacques Derrida, Northrop Frye, Harold Bloom, Geoffrey Hartman, Frank Kermode, Edward Said, Barbara Johnson, Frank Lentricchia, and J. Hillis Miller* (Methuen, 1987).
Sangari, Kumkum, 'The Politics of the Possible' in Ashcroft, Griffiths and Tiffin, 1995: 143–7.
Schleifer, R (ed.), *The Genres of the Irish Literary Revival* (Dublin: Wolfhound Press, 1980).
Schwarz, Bill (ed.), *The Expansion of England: Race, Ethnicity and Cultural History* (Routledge, 1996).
Sena, Vinod, *W.B. Yeats: The Poet as Critic* (Macmillan, 1980).
Shaw, Francis, 'The Canon of Irish History: A Challenge', *Studies: An Irish Quarterly Review of Letters, Philosophy and Science* 61 (Summer 1972), 113–52.
Slocum, John J. and Herbert Cahoon (eds), *A Bibliography of James Joyce* [1953] (Westport, Conn.: Greenwood Press, 1971).
Smith, Anthony D. *National Identity* (Penguin, 1991).
Smith, Paul, *Discerning the Subject* (Minneapolis: University of Minnesota Press, 1988).
Smyth, Gerry, '"The natural course of things": Matthew Arnold, Celticism and the English Poetic Tradition', *Journal of Victorian Culture* 1.1 (Spring 1996a), 35–53.
—— 'Being Difficult: The Irish Writer in Britain', *Éire–Ireland: An Interdisciplinary Journal of Irish Studies* (Fall/Winter 1996b), 41–57.
Snyder, E.D., *The Celtic Revival in English Literature 1760–1800* (Cambridge: Harvard University Press, 1923).
Sommerfelt, Alf (ed.), *A Review of Celtic Studies*, (Oslo: 1958).

Spivak, Gayatri Chakravorty, *In Other Worlds: Essays in Cultural Politics* (Methuen, 1987).
—— *The Post-Colonial Critic: Interviews, Strategies, Dialogues*, ed. Sarah Harasym, (Routledge, 1990).
Sprinker, Michael (ed.), *Edward Said: A Critical Reader* (Oxford: Blackwell, 1992).
Stanford, W.B., *The Ulysses Theme: A Study in the Adaptability of a Traditional Hero* [1954] (Oxford: Basil Blackwell, 1968).
Storey, Mark (ed.), *Poetry and Ireland since 1800: A Source Book* (Routledge, 1988).
Taylor, Geoffrey (ed.), *Irish Poets of the Nineteenth Century* (Routledge and Kegan Paul, 1951).
Tindall, William York, *James Joyce: His Way of Interpreting the World* (New York: Scribner's, 1950).
Todorov, Tzvetan, *On Human Diversity: Nationalism, Racism, and Exoticism in French Thought* [1989], trans. C. Porter (Harvard University Press, 1993).
Ussher, Arland, *The Face and Mind of Ireland* (Gollancz, 1949).
—— *Three Great Irishmen: Shaw, Yeats, Joyce* [1953] (New York: Biblo and Tannen, 1968).
Vance, Norman, *Irish Literature: A Social History* (Oxford: Basil Blackwell, 1990).
Walker, J.C., *Historical Memoirs of the Irish Bards; Interspersed with Anecdotes of, and Occasional Observations on The Music of Ireland; also, an Historical and Descriptive Account of the Musical Instruments of the Ancient Irish* [1786] (Dublin: J. Christie, 1818).
Watson, George, *The Literary Critics: A Study of English Descriptive Criticism* [1962] (Harmondsworth, Middlesex: Penguin, 1964).
Welch, Robert (ed.), *The Oxford Companion to Irish Literature* (Oxford: Clarendon Press, 1996).
Whelan, Bernadette, 'An Essay on Ireland and J. William Fulbright's Educational Vision', *Éire–Ireland: An Interdisciplinary Journal of Irish Studies* (Fall/Winter 1996), 153–75.
Whitaker, T.K., *Economic Development* (Dublin: The Stationery Office, 1958).
Wilde, Oscar, *Intentions and The Soul of Man under Socialism* [1891] (Dawsons of Pall Mall, 1969).
Williams, Patrick and Laura Chrisman (eds), *Colonial Discourse and Post-Colonial Theory: A Reader* (Hemel Hempstead: Harvester Wheatsheaf, 1993).
Williams, Raymond, *The Long Revolution* [1961] (Harmondsworth, Middlesex: Pelican, 1984).
—— *Keywords* [1976] (2nd edn 1983; Fontana, 1984).

Wilson, F.A.C., *W.B. Yeats and Tradition* (Methuen, 1958).
Wimsatt Jr., William K. and Cleanth Brooks, *Literary Criticism: A Short History* (New York: Alfred A. Knopf, 1957).
Yeats, W.B., 'Nationality and Literature' in Storey 1988: 85–92.
—— *The Collected Poems of W.B. Yeats* [1933] (Macmillan, 1981).
—— *Essays and Introductions* (Macmillan, 1961).
Young, Robert (ed.), *Untying the Text: A Post-Structuralist Reader* (Routledge & Kegan Paul, 1981).
Young, Robert J. C., *White Mythologies: Writing History and the West* (Routledge, 1990).
—— *Colonial Desire: Hybridity in Theory, Culture and Race* (Routledge, 1995).

Periodicals, Calendars, Official Publications

ASLIB Index to Theses Accepted for Higher Degrees in the Universities of Great Britain and Ireland, vol.1 (1950/51; London, 1953).
The Bell: A Literary Magazine (edited by Seán O'Faoláin and Peadar O'Donnell; Dublin, 1940–48, 1950–54).
Beltaine (edited by W.B.Yeats; May 1899–April 1900).
Celtica: the Journal of the School of Celtic Studies (Dublin 1946–).
The Dublin Magazine: A Quarterly Review of Literature, Science and Art (edited by Seamus O'Sullivan; Dublin, 2 series, 1923–25, 1926–58).
The Dublin University Calendar (Dublin: Hodges, Figgis & Co. Ltd).
Éigse: A Journal of Irish Studies (Dublin, 1939–).
Ériu: Founded as the Journal of the School of Irish Learning Devoted to Irish Philology and Literature (Dublin, 1904–).
Envoy: An Irish Review of Literature and Art (edited by John Ryan and Valentin Iremonger; Dublin, 1949–51).
The Furrow (edited by J.G. McGarry; Naas, 1949–).
Hermathena: A Series of Papers on Literature, Science and Philosophy by Members of Trinity College, Dublin (Dublin, 1913–).
The Irish Book (edited by Alf MacLochlann; Dublin, 1959–64).
The Irish Book Lover: A Monthly Review of Irish Literature and Bibliography (edited by John S. Crone, later by Seamus Ó Casaide, then Colm O'Lochlann; Dublin, 1909–57; superseded by *The Irish Book*).
The Irish Bookman (edited by Seamus Cambell; Dublin, 1956).

Irish Historical Studies: The Joint Journal of the Irish Historical Society and the Ulster Society for Irish Historical Studies (Dublin and Oxford).

Irish Writing: The Magazine of Contemporary Irish Literature (edited by David Marcus and Terence Smith, Cork, 1946–54; and by Seán J. White, 1954–57).

Kavanagh's Weekly: A Journal of Literature and Politics (edited by Patrick and Peter Kavanagh; nos 1–13, Dublin, 12 April–5 July, 1952).

Poetry Ireland (edited by David Marcus, Cork, 1948–54; John Jordan, Cork, 1963–68; superseded by *Poetry Ireland Review*, edited by John F. Deane *et al.*, Dublin, 1981–).

Rann: An Ulster Quarterly of Poetry and Comment (edited by Barbara Hunter and Roy McFadden; Belfast, 1948–53).

Register of Prohibited Publications: I. Supplementary list of Prohibited Books: II. Register of Prohibited Periodical Publications (Dublin: The Stationery Office, 1958).

Report of the Commission on Emigration and Other Population Problems, 1948–1954 (Dublin: The Stationery Office, 1956).

Studies: An Irish Quarterly Review of Letters, Philosophy and Science (Dublin, 1912–).

Threshold (edited by Mary O'Malley and John Hewitt; Belfast, 1957–).

University College Dublin: Calendar for the Session 1948–58 (Dublin: Browne and Nolan).

University Review: Official Organ of the Graduates Association of the National University of Ireland (edited by Lorna Reynolds; Dublin, 1954–68).

Index

Printed and bound by CPI Group (UK) Ltd, Croydon, CR0 4YY

27/10/2024

14580224-0002